Developmental Biology

Developmental Biology

Laila Prady

Larsen & Keller
www.larsen-keller.com

Developmental Biology
Laila Prady
ISBN: 978-1-64172-616-0 (Hardback)

Larsen & Keller
Published by Larsen and Keller Education,
5 Penn Plaza,
19th Floor,
New York, NY 10001, USA

Cataloging-in-Publication Data

Developmental biology / Laila Prady.
p. cm.
Includes bibliographical references and index.
ISBN 978-1-64172-616-0
1. Developmental biology. 2. Growth. 3. Ontogeny. 4. Biology. I. Prady, Laila.
QH491 .D48 2022
571.8--dc23

Table of Contents

Preface

The study of the processes that enable animals and plants to grow and develop is referred to as developmental biology. It encompasses various fields of biology such as the biology of regeneration, asexual reproduction and metamorphosis. This field is also concerned with the growth of stem cells in adult organisms. It focuses on the main processes involved in the embryonic development of animals such as cell differentiation, regional specification, morphogenesis and growth. Cell differentiation refers to the formation of different functional cell types like muscle, secretory epithelia and nerve. This textbook is a valuable compilation of topics, ranging from the basic to the most complex theories and principles in the field of developmental biology. Some of the diverse topics covered herein address the varied branches that fall under this category. Those in search of information to further their knowledge will be greatly assisted by this book.

Given below is the chapter wise description of the book:

Chapter 1- Developmental biology is the field of biology that studies the process by which animals and plants grow and develop. Early developmental biological processes include oogenesis, gametogenesis, spermatogenesis and fertilization. This is an introductory chapter which will introduce briefly all these significant aspects of developmental biology.

Chapter 2- There are various approaches towards developmental biology such as anatomical approaches, experimental approaches and genetic approaches. This chapter closely examines these approaches towards developmental biology as well as key concepts related to it such as cell division, gene cloning and genetic mosaics.

Chapter 3- The phase of embryonic development that starts at the end of gastrulation and lasts till birth is known as organogenesis. This chapter has been carefully written to provide an easy understanding of the different processes related to development of organs such as organogenesis and cellular differentiation.

Chapter 4- The process by which structures originate and mature as a plant grows is known as plant development. Plants continuously produce new structures and tissues from a tissue known as meristem. The topics elaborated in this chapter will help in gaining a better perspective about the processes related to plant growth and development as well as its stages.

Chapter 5- The processes that lead eventually to the formation of a new animal starting from cells derived from one or more parent individuals is known as animal development. Some of the stages of animal development are embryo formation, embryonic adaptations and organ formation. This chapter closely examines these key concepts as well as the stages of animal development to provide an extensive understanding of the subject.

At the end, I would like to thank all those who dedicated their time and efforts for the successful completion of this book. I also wish to convey my gratitude towards my friends and family who supported me at every step.

Laila Prady

1

Understanding Developmental Biology

Developmental biology is the field of biology that studies the process by which animals and plants grow and develop. Early developmental biological processes include oogenesis, gametogenesis, spermatogenesis and fertilization. This is an introductory chapter which will introduce briefly all these significant aspects of developmental biology.

Developmental biology is the science of explaining how a variety of interacting processes generate an organism's heterogeneous shapes, size, and structural features that arise on the trajectory from embryo to adult, or more generally throughout a life cycle. It represents an exemplary area of contemporary experimental biology that focuses on phenomena that have puzzled natural philosophers and scientists for more than two millennia. In Generation of Animals, Aristotle provided the first systematic investigation of developmental phenomena and recognized key issues about the emergence of and relationships between hierarchically organized parts (e.g., bone and anatomical features containing bone), as well as the explanatory difficulty of determining how a morphological form is achieved reliably in offspring (e.g., the typical shape and structure of appendages in a particular species). Generation remained a poignant question throughout the early modern period, and was explored by many key figures writing at the time, including William Harvey, René Descartes, Robert Boyle, Pierre Gassendi, Nicolas Malebranche, Gottfried Wilhelm Leibniz, Anne Conway, Immanuel Kant, and others. Observations of life cycle transitions, such as metamorphosis, fed into these endeavors and led to striking conclusions, such as Leibniz's denial of generation sensu stricto.

Animals and all other organized substances have no beginning their apparent generation is only a development, a kind of augmentation, a transformation like any other, for instance like that of a caterpillar into a butterfly.

A major theme that crystallized in this history of investigation is the distinction between epigenesis and preformation. Proponents of epigenesis claimed that heterogeneous, complex features of form emerge from homogeneous, less complex embryonic structures through interactive processes. Thus, an explanation of the ontogeny of these form features requires accounting for how the interactions occur. Proponents of preformation claimed that complex form preexists in the embryo and "unfolds" via ordinary growth processes. An adequate explanation involves detailing how growth occurs. Although preformation has a lighter explanatory burden in accounting for how form emerges during ontogeny (on the assumption that growth is easier to explain than process interactions), it also must address how the starting point of the next generation is formed with the requisite heterogeneous complex features. This was sometimes accomplished by embedding smaller and smaller miniatures ad infinitum inside the organism. Epigenetic perspectives were often dependent on forms of teleological reasoning to account for why interactions among homogeneous components eventually result in a complex, integrated whole organism. Though nothing prevents mixing features of these two outlooks in explaining different aspects of development, polarization into dichotomous positions has occurred frequently.

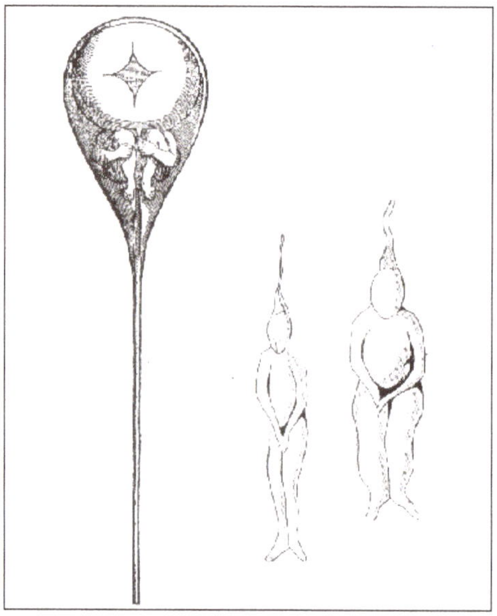

An early modern depiction of a tiny person inside of a sperm exemplifying preformationist views.

In the late 19th and early 20th century, the topic of development was salient in controversies surrounding vitalism, such as the disagreement between Wilhelm Roux and Hans Driesch over how to explain ontogeny. Roux thought that a fertilized egg contains inherited elements that represent different organismal characteristics. During the process of cellular division, these elements become unequally distributed among daughter cells leading to distinct cell fates. Driesch, in contrast, held that each cell retained its full potential through division even though differentiation occurred. Although this issue is often understood in terms of the metaphysics of life (vitalism versus materialism), Driesch's interpretation of development and the autonomy of an organism had epistemological dimensions. The explanatory disagreement involved different experimental approaches and divergent views on the nature of differentiation in early ontogeny (e.g., to what degree cells are pre-specified). A familiar philosophical theme running through these discussions, both epistemological and metaphysical, is the status of reductionism in biology. Through the middle of the 20th century, embryology—the scientific discipline studying development—slowly transformed into developmental biology with a variety of reworked and recalcitrant elements. In conjunction with the issue of reductionism, a key aspect of this history is the molecularization of experimental (as opposed to comparative) embryology, with the concomitant stress on the explanatory power of genes. This complex and fascinating history has been detailed elsewhere; Embryo Project in Other Internet Resources).

Developmental biology has not been a primary area of exploration for philosophy of biology though this has been changing over the past two decades, especially due to the potential relevance of development for understanding evolution, the theme of reductionism in biology and explanations from molecular genetics, and via increased attention to the details of particular research programs, such as stem cell biology. But it should not be forgotten that ontogeny was on the radar of philosophical scholars in the 20th century, as seen in Ernest Nagel's treatment of hierarchical organization and reduction in the development of living systems. For contemporary philosophy of science, developmental biology displays a rich array of material and conceptual practices that can be analyzed to better understand the scientific reasoning exhibited in experimental life science.

After a brief review of some central phenomena of ontogeny, this entry explores four domains that compose a subset of the import and promise of conceptual reflection on the epistemology of developmental biology.

Developmental Phenomena

Developmental biology is the science that seeks to explain how the structure of organisms changes with time. Structure, which may also be called morphology or anatomy, encompasses the arrangement of parts, the number of parts, and the different types of parts.

Most of the properties that developmental biologists attempt to explain are structural rather than functional. For example, a developmental biologist concentrates on how tissue layers fold or how shape is generated, and not so much on what the folded tissue layers do or how the shape functions. The ontogeny of function, at all levels of organization, is an element of developmental biology, but it is often bracketed because of the predominance (both past and present) of questions surrounding the ontogeny of form or structure.

Textbooks typically describe a canonical set of events surrounding the changing structures displayed during animal development. The first of these is fertilization (in sexually reproducing species), where an already semi-organized egg merges with a sperm cell, followed by the fusion of their nuclei to achieve the appropriate complement of genetic material. Second, the fertilized egg undergoes several rounds of cleavage, which are mitotic divisions without cell growth that subdivide the zygote into many distinct cells. After many rounds of cleavage, this spherical conglomerate of cells (now called a blastula) begins to exhibit some specification of germ layers (endoderm, mesoderm, and ectoderm) and then proceeds to invaginate at one end, a complex process referred to as gastrulation that eventually yields a through-gut. All three germ layers, from which specific types of cells are derived (e.g., neural cells from ectoderm), become established during gastrulation or shortly after it completes. Organogenesis refers to the production of tissues and organs through the interaction and rearrangement of cell groups. Events confined to distinct taxonomic groups include neurulation in chordates, whereas other events correlate with mode of development (metamorphosis from a larval to adult stage) or individual trauma (regeneration of a limb).

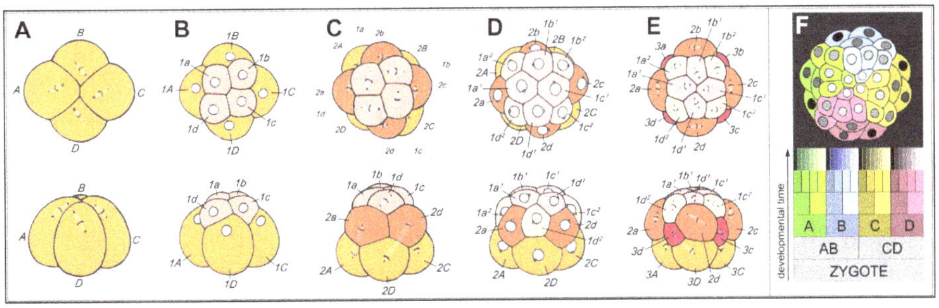

An example of embryonic cleavage in marine snail embryos showing the fate of different cell lineages through developmental time.

Several key processes underlie these distinct developmental events and the resulting features of form that emerge (e.g., the through-gut formed subsequent to gastrulation or the heart formed during organogenesis). These are critical to the ontogeny of form and link directly to major research questions in developmental biology. First, cellular properties, such as shape, change during ontogeny. This is a function of differentiation whereby cells adopt specific fates that include shape transformations. Second, regions of cells in the embryo are designated through arrangement and composition alterations that correspond to different axes in different parts of the embryo (e.g., dorsal-ventral, anterior-posterior, left-right, and proximal-distal). The successive establishment of these regions is referred to as pattern formation. Third, cells translocate and aggregate into layers (e.g., endoderm and ectoderm, followed by the mesoderm in many lineages) and later tissues (aggregations of differentiated cell types). Fourth, cells and tissues migrate and interact to produce new arrangements and shapes composed of multiple tissue layers with novel functions (i.e., organs). These last two sets of processes are usually termed morphogenesis and occur via many distinct mechanisms. Fifth, there is growth in the size of different form features in the individual, remarkably obvious when comparing zygote to adult, although proportional change between different features (allometry) is also striking.

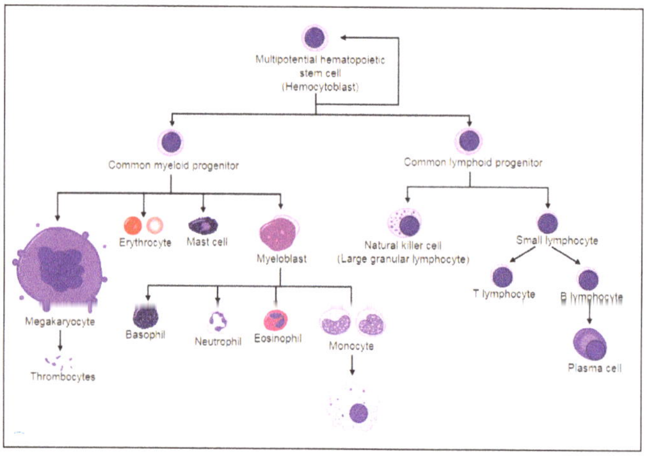

A simple illustration of the kinds of differentiation related to the cellular components found in blood.

None of these processes occur in isolation and explanations of particular form features usually draw on several of them simultaneously, presuming other features that originated earlier in ontogeny by different instantiations and combinations of the processes. This sets a broad agenda for investigation: how do various iterations and combinations of these processes generate form features during ontogeny? This can be seen more concretely through the example of vertebrate cardiogenesis. How does the vertebrate heart, with its internal and external shape and structure originate during ontogeny? How does the heart come to exhibit left/right asymmetry in the body cavity? What causes cells to adopt a muscle cell fate or certain tissues to interact in the prospective region of the heart? How do muscle cells migrate to, aggregate in, and differentiate at the correct location? How does the interior of the heart adopt a particular tubular structure with various chambers (that differs among vertebrate species)? How does the heart grow at a particular rate and achieve a specific size? Solutions relevant to explaining the ontogeny of form characterize causal factors that account for how different processes occur and yield various outcomes.

The Epistemological Organization of Developmental Biology

One recurring theme in the long history of investigations into development is that explaining the ontogeny of form consists of many interrelated questions about diverse phenomena. Sometimes philosophers have attempted to compress these questions into one broad problem.

The real question concerning metazoan ontogeny is just how a single cell gives rise to the requisite number of differentiated cell lineages with all the right inductive developmental interactions required to reproduce the form of the mature organism.

The central problem of developmental biology is to understand how a relatively simple and homogeneous cellular mass can differentiate into a relatively complex and heterogeneous organism closely resembling its progenitor(s) in relevant aspects.

This language is not necessarily incorrect but can lead to skewed interpretations. For example, Philip Kitcher has argued that:

> "In contemporary developmental biology, there is uncertainty about how to focus the big, vague question, how do organisms develop."

This is simply false. While it is true that these questions have been manifested with differing frequency and vigor through history and the ability to answer them (as well as the nature of the questions themselves) has been contingent on different research strategies and methods, the issue has not been an unwieldy central problem. But scrutinizing the structure of developmental biology's questions is not merely an exercise in clarification. It is crucial for understanding how the science of developmental biology is organized.

Erotetic Organization

The criterion of epistemic transparency encourages an exploration of the third interpretive option—the lack of reference to theories should be taken at face value. Developmental biology is organized primarily by stable, broad domains of problems that correspond to abstract representations of major ontogenetic processes (differentiation, pattern formation, growth, and morphogenesis). But how does one interpret the "theoretical" aspects of developmental biology (e.g., positional information models of pattern formation) and the utilization of theories from other domains (e.g., biochemistry)? One way is to distinguish between theory-informed science—using theoretical knowledge—and theory-directed science—having a theory that directs inquiry and organizes knowledge; developmental biology is theory-informed but not theory-directed. Theories need not be wholly absent from developmental biology but—when present—they play roles very different from standard philosophical expectations. Developmental biology uses theoretical knowledge from biochemistry when appealing to morphogen gradients to explain how segments are established or chemical thermodynamics when invoking reaction–diffusion mechanisms to explain pigmentation patterns. It also uses theoretical knowledge derived from within developmental biology, such as positional information models. Different kinds of theory inform developmental biology, but these do not organize research—they are not necessary to structure the knowledge and direct investigative activities. Developmental biologists are not focused on confirming and extending the theory of reaction–diffusion mechanisms, nor are they organizing their research around positional information. This theoretical knowledge is used in building explanations but does not provide rails of guidance for how to proceed in a research program. All sciences may use theoretical knowledge, but this is not the same as all sciences having a theory providing direction and organization.

Why think that problems provide organizational architecture for the epistemology of developmental biology? They are a pervasive feature of its reasoning practices, illustrated in textbooks that capture substantial community consensus about standards of explanation, experimental methods, essential concepts, and empirical content. Unlike evolutionary biology textbooks that discuss the theory of natural selection or economics textbooks that talk about microeconomic theory, an examination of several editions of major textbooks indicate that developmental biology does not have similar kinds of theories.

Jonathan Slack's Essential Developmental Biology is organized around four main types of processes, also described as clustered groups of problems, which occur during embryonic development: regional specification (pattern formation), cell differentiation, morphogenesis, and growth. These broad clusters are then fleshed out along a standard timeline of early development, highlighting gametogenesis, fertilization, cleavage, gastrulation, and axis specification. Different experimental approaches (cell and molecular biology, developmental genetics, and experimental embryology) are utilized in a specific set of model organisms to dissect the workings of these developmental phenomena. Subsequent chapters cover later aspects of development (e.g., organogenesis), with different systems treated in depth by tissue layer, differentiation and growth, or in relation to evolutionary questions. Throughout this presentation, no specific theory, set of hypotheses, or dominant model is invoked to organize these different domains of investigation. Instead, broad clusters of questions that reflect generally delineated processes (differentiation, specification, morphogenesis, and growth) set the agenda of research.

Scott Gilbert's Developmental Biology exhibits a similar pattern. Developmental biology is constituted by two broad questions ("How does the fertilized egg give rise to the adult body? And how does that adult body produce yet another body?"), which can then be subdivided into further categories, such as differentiation, morphogenesis, growth, reproduction, regeneration, evolution, and environmental regulation. These questions can be parsed more analytically in terms of five variables: abstraction, variety, connectivity, temporality, and spatial composition. The values given to these variables structure the constellation of research questions within the broad problem agendas corresponding to generally delineated processes. For example, research questions oriented around events in zebrafish gastrulation are structured in a way that differs from the research questions oriented around vertebrate neural crest cell migration because they involve different values for the five variables: abstraction (zebrafish vs. vertebrates), temporality (earlier vs. later), spatial composition (tissue layer interactions vs. a distinctive population of cells), variety (epiboly vs. epithelium to mesenchyme transition), and connectivity (gut formation and endoderm vs. organogenesis and ectoderm/mesoderm). These configurations can be adjusted readily in response to shifts in the values for different variables.

This anatomy of problems, with explicit epistemological structure derived from different values for these variables, operates to organize the science of development. Investigators from different disciplines can be working on the same problem but asking different questions that require distinct but complementary methodological resources. Knowledge and inquiry in developmental biology are intricately organized, just not by a central theory or group of models, and this erotetic organization is epistemologically accessible to the participating scientists. While theoretical knowledge, especially that drawn from molecular biological mechanisms and mathematical models (e.g., reaction–diffusion models) is ubiquitous (theory-informed), the clusters of problems that reappear across the textbooks and correspond to different types of processes provide the governing architecture (not theory-directed), which can be characterized explicitly according to the variables described. Further analysis of this problem anatomy is possible, including how it is displayed in regular research articles and not just textbooks, as well as other areas of biology.

Explanatory Approaches to Development

Explanations in developmental biology are usually causal, though unlike standard mechanistic explanation there is a constant acquisition of new causal capacities (in terms of constituent entities, activities, and their organization) through development. Although much work remains in characterizing different aspects of explanation in developmental biology, there is no doubt that a difference making or manipulability conception of causation provides a core element of the reasoning. Genetic explanations of development, similar to what is seen in molecular genetics, work by identifying changes in the expression of genes and interactions among their RNA and protein products that lead to changes in the properties of morphological features during ontogeny (e.g., shape or size), while holding a variety of contextual variables fixed. More recently, there has been growing interest in physical explanations of development that involve appeals to mechanical forces due to the geometrical arrangements of mesoscale materials, such as fluid flow. Researchers agree on the phenomena that need to be explained, but differ on whether physical rules or genetic factors are more or less explanatory. The existence of two different types of causal explanations for developmental phenomena poses an additional question about how they might (or might not) be combined into a more integrated explanatory framework.

Genetics

Many philosophers have turned to explanations of development over the past two decades in an effort to esteem or deflate claims about the causal power of genes. Genetic explanations touch the philosophical theme of reductionism and appear to constitute the bulk of empirical success accruing to developmental biology over the past several decades. Statements from developmental biologists reinforce this perspective:

> "Developmental biology deals with the process by which the genes in the fertilized egg control cell behavior in the embryo and so determine its pattern, its form, and much of its behavior differential gene activity controls development."

These types of statements are amplified in appeals to a genetic program for development.

Elements of the genome contain the sequence-specific code for development; and they determine the particular outcome of developmental processes, and thus the form of the animal produced by every embryo. Development is the execution of the genetic program for construction of a given species of organism.

Whether or not these statements can be substantiated has been the subject of intense debate. The strongest claims about genetic programs or the genetic control of development have empirical and conceptual drawbacks that include an inattention to plasticity and the role of the environment, an ambiguity about the locus of causal agency, and a reliance on metaphors drawn from computer science. But this leaves intact the difference-making principle of genetic explanation exhibited in molecular genetics, which yields more narrow and surgical causal claims under controlled experimental conditions, and is applicable to diverse molecular entities that play causal roles during development, such as regulatory RNAs, proteins, and environmental signals. We can observe this by reconsidering the example of vertebrate cardiogenesis.

Are there problems with claiming that genes contain all of the information to form vertebrate hearts? Is there a genetic program in the DNA controlling heart development? Are genes the primary supplier and organizer of material resources for heart development, largely determining the phenotypic outcome? Existing studies of heart development have identified a role for fluid forces in specifying the internal form of the heart and its left/right asymmetry. Biochemical gradients of extracellular calcium are responsible for activating the asymmetric expression of the regulatory gene Nodal and inhibition of voltage gradients scrambles normal asymmetry establishment. A number of genes are clearly difference makers in these processes, but the conclusion that genes carry all the information needed to generate form features of the heart is unnecessary and unwarranted. While it may be warranted empirically in some cases to privilege DNA sequence differences as causal factors in specific processes of ontogeny, such as hierarchically organized networks of genetic difference makers explaining tissue specification, the diversity of entities appealed to in molecular genetics and the extent of their individual and joint roles in specifying developmental outcomes means that debates about the meaning, scope, and power of genetic explanations will continue. But a shift away from genetic programs and genetic determinism to DNA, RNA, and proteins as difference makers that operate conjointly suggests that we conceptualize other causal factors in a similar way.

Physics

Fluid flow, as a physical force, is also a difference maker during the development of the heart, and ontogeny more generally, and developmental biologists appeal to physical difference makers, which are understood as factors in producing the morphological properties of developmental phenomena. A physical causation approach was displayed prominently in the first half of the 20th century through the work of D'Arcy Thompson. This occurred in the milieu of increasing attention to the chromosomal theory of inheritance and attempts to explore developmental phenomena via classical genetic methods. Thompson appealed to differential rates of growth and the constraints of geometrical relationships to explain how organismal morphology originates. Visual representations of abiotic, mechanical analogues provided the plausibility, such as the shape of liquid splashes or hanging drops for the cup and bell configurations of the free-swimming sexual stage of jellyfish. If physical forces generated specific morphologies in viscoelastic materials, then analogous morphologies in living species should be explained in terms of physical forces operating on the viscoelastic materials of the developing embryo. But morphogenetic processes that produce the shape and structure of morphology have been seen primarily, if not exclusively, in terms of genetics for the last half-century. Physical approaches faded from view as molecular genetics approaches went from strength to strength.

The molecularization of experimental embryology is one of the most striking success stories of contemporary biology as genes and genetic interactions (e.g., in transcriptional networks and signaling pathways) were discovered to underlie specific details of differentiation, morphogenesis, pattern formation, and growth when structure originates during development. Genetic approaches predominate in contemporary developmental biology; physical modes of causation are typically neglected, if not negligible. The frustration among researchers interested in physical causation during embryogenesis has been palpable.

To the molecular types, a cause is a molecule or a gene. To explain a phenomenon is to identify genes and characterize proteins without which the phenomenon will fail or be abnormal. A molecule is an explanation: a force is a description; to argue otherwise brings pity, at best.

Despite this predominance of genetic explanatory approaches and the frustration among researchers utilizing other approaches, a groundswell of interest has been building around physical explanations of development, especially in terms of their integration with genetic explanations.

Integrating Approaches: Genetics and Physics

Thompson held that physical forces were explanatory but inadequate in isolation to account for the developmental origin of morphology; heredity (genetics) was also a necessary causal factor. Yet Thompson was quick to highlight that mechanical modes of causation might be neglected in the midst of growing attention to heredity (genetics):

> "It is no less of an exaggeration if we tend to neglect these direct physical and mechanical modes of causation altogether, and to see in the characters of a bone merely the results of variation and of heredity."

Despite this latter form of exaggeration manifesting itself through much of the 20th century, an agenda to combine or integrate the two approaches is now explicit.

There is no controversy about whether genetic and physical modes of causation are at work simultaneously:

> "Both the physics and biochemical signaling pathways of the embryo contribute to the form of the organism. "

They are not competing causal explanations of the same phenomenon. Explanations should capture how their productive interactions yield developmental outcomes:

> "An increasing number of examples point to the existence of a reciprocal interplay between expression of some developmental genes and the mechanical forces that are associated with morphogenetic movements."

Genetic causes can lead to physical causation and vice versa. Physical causation brings about genetic causation through mechanotransduction. Stretching, contraction, compression, fluid shear stress, and other physical dynamics are sensed by different molecular components inside and outside of cells that translate these environmental changes into biochemical signals. Genetic causation brings about physical causation by creating different physical properties of cells and tissues through the presence, absence, or change in frequency of particular proteins. For example, different patterns of expression for cell adhesion molecules (e.g., cadherins) can lead to differential adhesion across epithelial sheets of tissue and thereby generate phase separations or compartments via surface tension variation. If these modes of causation are not competing, then how might one combine genetic and physical difference makers into an integrated causal explanation? How much explanatory unity can be achieved for this "reciprocal interplay."

Finding philosophical models for the explanatory integration of genetics and physics remains an open question (Love forthcoming). Apportioning causal responsibility in the sense of determining relative contributions (e.g., the composition of causal magnitudes among different physical forces in Newtonian mechanics) is problematic because this requires commensurability with respect to how causes produce their effects. In the context of causation understood in terms of difference makers, the difficulty of integration is a variation on a problem in causal reasoning identified by John Stuart Mill and labeled the "intermixture of effects," which involves multiple causes contributing in a blended fashion to yield an outcome.

This difficulty is most of all conspicuous in the case of physiological phenomena; it being seldom possible to separate the different agencies which collectively compose an organized body, without destroying the very phenomena which it is our object to investigate.

Careful statistical methodology in experiments can answer whether one type of difference maker accounts for more of the variation in the effect variable for a particular population. But a ranking of causal factors with respect to how much of a difference they made is not the same as combining two modes of causation into an integrated account. Another response is to dissolve the integration problem by reducing all of the causal interactions to one of the two distinct modes, thereby achieved a kind of explanatory unity. But this approach is eschewed by working biologists who take both genetic and physical modes of causation as significant and not reducible one to the other.

A different strategy is integrative pluralism. This involves a two-step procedure for explaining complex phenomena whose features are the result of multiple causes: (a) formulate idealized models where particular causal factors operate in isolation ("theoretical modeling") and (b) integrate idealized models to explain how particular, concrete phenomena originate from these causes in combination. This model is suggestive but also has key drawbacks that include the fact that genetic causal reasoning in developmental biology does not involve formulating idealized models and the precise nature of the integration is underspecified. Integration of genetic and physical difference makers in a single mechanism offers a further possibility. Although this valuably highlights the productive continuity between difference makers through stages in a sequence (i.e., their reciprocal interplay), it also has handicaps. These include:

1. Divergent approaches to measuring time. Instead of time "in the mechanism," time is measured with external standardized stages. Stages facilitate the study of different kinds of developmental mechanisms, with different characteristic rates and durations for their stages, within a common framework for a model organism (e.g., Drosophila), while also permitting conserved molecular mechanisms to be studied in different species because the corresponding mechanism description is not anchored to the temporal sequence of the model organism.

2. An expectation that mechanism descriptions "bottom-out" in lowest level activities of molecular entities. In the case of combining genetic and physical difference makers, the reciprocal interplay means that there is a studious avoidance of bottoming out in one or the other mode of causation.

3. The requirement of stable, compositional organization for mechanisms: Mechanistic explanations are constitutive or componential explanations: they explain the behavior of the mechanism as a whole in terms of the organized activities and interaction of its components.

But these mechanism descriptions are often embedded in different developmental contexts (at different times in ontogeny) with distinct compositional relations (within and between species). The reciprocal interplay between genetic and physical difference makers is not maintained precisely because these compositional differences alter relationships of physical causation (fluid flow, tension, etc.). Developmental biologists have been able to generalize relationships of genetic causation (in terms of genetic mechanisms) across species quite widely, but the attempt to combine these with physical causation has necessitated narrowing the scope of the causal claims.

Adequate philosophical models of the systematic dependence between genetic and physical difference makers in ontogeny need to account for how the temporal relations necessary for making causal claims are anchored in an external periodization used by developmental biologists. The imposition of different temporal scales can lead to distinct factors being significant or salient, which matters for ascertaining how different kinds of causes can be combined into integrated explanations. One possibility is to juxtapose these difference makers at distinct stages via experimental verification such that they exhibit productive continuity within the constraints of the external periodization (Love forthcoming). This facilitates representing symmetry between causal factors because genetic difference makers can be placed before or after physical difference makers (and vice versa). Although this does not provide a way to combine causal magnitudes (as in vector addition from Newtonian mechanics), it offers an explicit strategy for assigning responsibility among different kinds of causes through the vehicle of temporal organization that goes beyond ranking difference makers. The periodization serves as a template from the practices of developmental biologists for providing wholeness or unity to the different modes of causation to yield a kind of integrated explanation of the morphology that result from a sequence of developmental processes.

Not all types of causal explanation involve an external periodization and there are other ways to combine causes in order to produce more integrated explanatory frameworks. One area where combined explanations for developmental phenomena are being analyzed pertains to mechanism descriptions and mathematical modeling in systems biology. For example, Fagan shows how an integrated explanation emerges from a step-wise procedure that starts with a detailed description of a molecular mechanism followed by the formulation of an abstracted wiring diagram of component interactions, which is then translated into a system of equations that can account for changes in component interactions over time. Solutions to these systems of equations and a mapping of solutions for the interactions among components of the system onto the behavior of the overall system within a shared landscape representation more systematically explains cellular differentiation.

Developmental Biology

The main processes involved in the embryonic development of animals are: regional specification, morphogenesis, cell differentiation, growth, and the overall control of timing explored in evolutionary developmental biology:

- Regional specification refers to the processes that create spatial pattern in a ball or sheet of initially similar cells. This generally involves the action of cytoplasmic determinants, located within parts of the fertilized egg, and of inductive signals emitted from signaling centers in the embryo. The early stages of regional specification do not generate functional differentiated cells, but cell populations committed to develop to a specific region or part of the organism. These are defined by the expression of specific combinations of transcription factors.

- Morphogenesis relates to the formation of three-dimensional shape. It mainly involves the orchestrated movements of cell sheets and of individual cells. Morphogenesis is important for creating the three germ layers of the early embryo (ectoderm, mesoderm and endoderm) and for building up complex structures during organ development.
- Cell differentiation relates specifically to the formation of functional cell types such as nerve, muscle, secretory epithelia etc. Differentiated cells contain large amounts of specific proteins associated with the cell function.
- Growth involves both an overall increase in size, and also the differential growth of parts (allometry) which contributes to morphogenesis. Growth mostly occurs through cell division but also through changes of cell size and the deposition of extracellular materials.
- The control of timing of events and the integration of the various processes with one another is the least well understood area of the subject. It remains unclear whether animal embryos contain a master clock mechanism or not.

The development of plants involves similar processes to that of animals. However plant cells are mostly immotile so morphogenesis is achieved by differential growth, without cell movements. Also, the inductive signals and the genes involved are different from those that control animal development.

Developmental Processes

Cell Differentiation

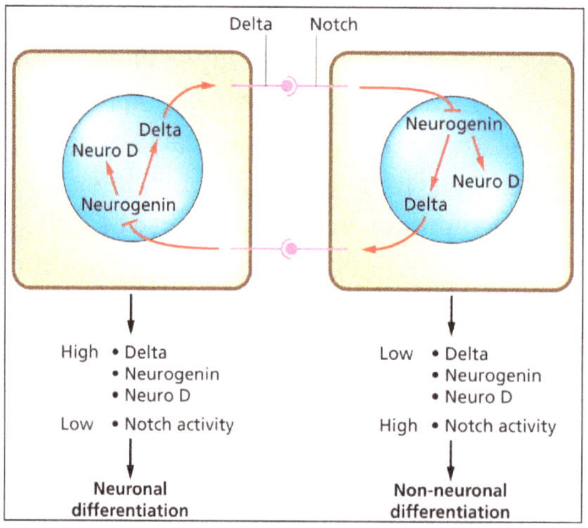

The Notch-delta system in neurogenesis.

Cell differentiation is the process whereby different functional cell types arise in development. For example, neurons, muscle fibers and hepatocytes (liver cells) are well known types of differentiated cells. Differentiated cells usually produce large amounts of a few proteins that are required for their specific function and this gives them the characteristic appearance that enables them to be recognized under the light microscope. The genes encoding these proteins are highly active. Typically their chromatin structure is very open, allowing access for the transcription enzymes, and specific transcription factors bind to regulatory sequences in the DNA in order to activate gene expression. For example, NeuroD is a key transcription factor for neuronal differentiation, myogenin for muscle differentiation, and HNF4 for hepatocyte differentiation.

Cell differentiation is usually the final stage of development, preceded by several states of commitment which are not visibly differentiated. A single tissue, formed from a single type of progenitor cell or stem cell, often consists of several differentiated cell types. Control of their formation involves a process of lateral inhibition, based on the properties of the Notch signaling pathway. For example, in the neural plate of the embryo this system operates to generate a population of neuronal precursor cells in which NeuroD is highly expressed.

Regeneration

Regeneration indicates the ability to regrow a missing part. This is very prevalent amongst plants, which show continuous growth, and also among colonial animals such as hydroids and ascidians. But most interest by developmental biologists has been shown in the regeneration of parts in free living animals. In particular four models have been the subject of much investigation. Two of these have the ability to regenerate whole bodies: *Hydra*, which can regenerate any part of the polyp from a small fragment, and planarian worms, which can usually regenerate both heads and tails. Both of these examples have continuous cell turnover fed by stem cells and, at least in planaria, at least some of the stem cells have been shown to be pluripotent. The other two models show only distal regeneration of appendages. These are the insect appendages, usually the legs of hemimetabolous insects such as the cricket, and the limbs of urodele amphibians. Considerable information is now available about amphibian limb regeneration and it is known that each cell type regenerates itself, except for connective tissues where there is considerable interconversion between cartilage, dermis and tendons. In terms of the pattern of structures, this is controlled by a re-activation of signals active in the embryo. There is still debate about the old question of whether regeneration is a "pristine" or an "adaptive" property. If the former is the case, with improved knowledge, we might expect to be able to improve regenerative ability in humans. If the latter, then each instance of regeneration is presumed to have arisen by natural selection in circumstances particular to the species, so no general rules would be expected.

Embryonic Development of Animals

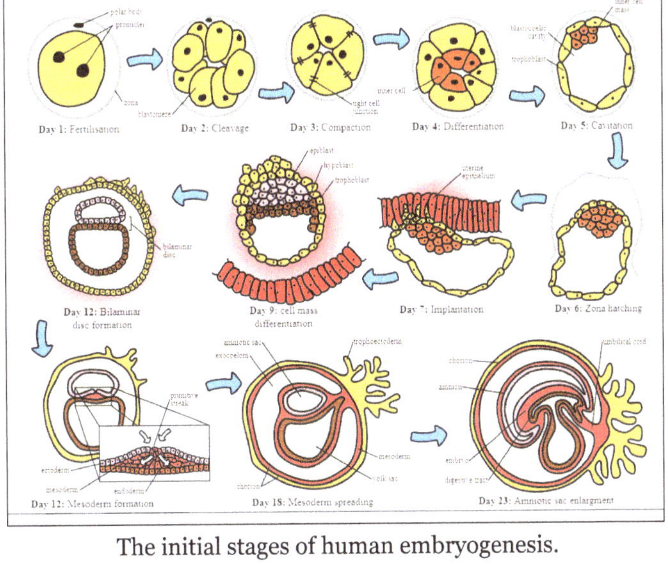

The initial stages of human embryogenesis.

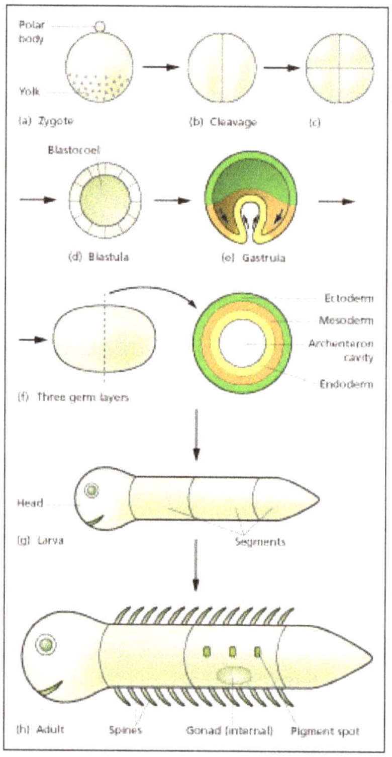

Generalized scheme of embryonic development: Slack "Essential Developmental Biology".

The sperm and egg fuse in the process of fertilization to form a fertilized egg, or zygote. This undergoes a period of divisions to form a ball or sheet of similar cells called a blastula or blastoderm. These cell divisions are usually rapid with no growth so the daughter cells are half the size of the mother cell and the whole embryo stays about the same size. They are called cleavage divisions.

Mouse epiblast primordial germ cells undergo extensive epigenetic reprogramming. This process involves genome-wide DNA demethylation, chromatin reorganization and epigenetic imprint erasure leading to totipotency. DNA demethylation is carried out by a process that utilizes the DNA base excision repair pathway.

Morphogenetic movements convert the cell mass into a three layered structure consisting of multicellular sheets called ectoderm, mesoderm and endoderm, which are known as germ layers. This is the process of gastrulation. During cleavage and gastrulation the first regional specification events occur. In addition to the formation of the three germ layers themselves, these often generate extraembryonic structures, such as the mammalian placenta, needed for support and nutrition of the embryo, and also establish differences of commitment along the anteroposterior axis (head, trunk and tail).

Regional specification is initiated by the presence of cytoplasmic determinants in one part of the zygote. The cells that contain the determinant become a signaling center and emit an inducing factor. Because the inducing factor is produced in one place, diffuses away, and decays, it forms a concentration gradient, high near the source cells and low further away. The remaining cells of the embryo, which do not contain the determinant, are competent to respond to different concentrations by upregulating specific developmental control genes. This results in a series of zones becoming set up, arranged at progressively greater distance from the signaling center. In each zone a different combination of developmental control genes is upregulated. These genes encode transcription factors which upregulate new combinations of gene activity in each region. Among other functions, these transcription factors control expression of genes conferring specific adhesive and motility properties on the cells in which they are active. Because of these different morphogenetic properties, the cells of each germ layer move to form sheets such that the ectoderm ends up on the outside, mesoderm in the middle, and endoderm on the inside. Morphogenetic movements not only change the shape and structure of the embryo, but by bringing cell sheets into new spatial relationships they also make possible new phases of signaling and response between them.

Growth in embryos is mostly autonomous. For each territory of cells the growth rate is controlled by the combination of genes that are active. Free-living embryos do not grow in mass as they have no external food supply. But embryos fed by a placenta or extraembryonic yolk supply can grow very fast, and changes to relative growth rate between parts in these organisms help to produce the final overall anatomy.

The whole process needs to be coordinated in time and how this is controlled is not understood. There may be a master clock able to communicate with all parts of the embryo that controls the course of events, or timing may depend simply on local causal sequences of events.

Metamorphosis

Developmental processes are very evident during the process of metamorphosis. This occurs in various types of animal. Well-known are the examples of the frog, which usually hatches as a tadpole and metamorphoses to an adult frog, and certain insects which hatch as a larva and then become remodeled to the adult form during a pupal stage.

All the developmental processes listed above occur during metamorphosis. Examples that have been especially well studied include tail loss and other changes in the tadpole of the frog *Xenopus*, and the biology of the imaginal discs, which generate the adult body parts of the fly *Drosophila melanogaster*.

Plant Development

Plant development is the process by which structures originate and mature as a plant grows. It is studied in plant anatomy and plant physiology as well as plant morphology.

Plants constantly produce new tissues and structures throughout their life from meristems located at the tips of organs, or between mature tissues. Thus, a living plant always has embryonic tissues. By contrast, an animal embryo will very early produce all of the body parts that it will ever have in its life. When the animal is born (or hatches from its egg), it has all its body parts and from that point will only grow larger and more mature.

The properties of organization seen in a plant are emergent properties which are more than the sum of the individual parts. "The assembly of these tissues and functions into an integrated multicellular organism yields not only the characteristics of the separate parts and processes but also quite a new set of characteristics which would not have been predictable on the basis of examination of the separate parts."

Growth

A vascular plant begins from a single celled zygote, formed by fertilisation of an egg cell by a sperm cell. From that point, it begins to divide to form a plant embryo through the process of embryogenesis. As this happens, the resulting cells will organize so that one end becomes the first root, while the other end forms the tip of the shoot. In seed plants, the embryo will develop one or more "seed leaves" (cotyledons). By the end of embryogenesis, the young plant will have all the parts necessary to begin in its life.

Once the embryo germinates from its seed or parent plant, it begins to produce additional organs (leaves, stems, and roots) through the process of organogenesis. New roots grow from root meristems located at the tip of the root, and new stems and leaves grow from shoot meristems located at the tip of the shoot. Branching occurs when small clumps of cells left behind by the meristem, and which have not yet undergone cellular differentiation to form a specialized tissue, begin to grow as the tip of a new root or shoot. Growth from any such meristem at the tip of a root or shoot is termed primary growth and results in the lengthening of that root or shoot. Secondary growth results in widening of a root or shoot from divisions of cells in a cambium.

In addition to growth by cell division, a plant may grow through cell elongation. This occurs when individual cells or groups of cells grow longer. Not all plant cells will grow to the same length. When cells on one side of a stem grow longer and faster than cells on the other side, the stem will bend to the side of the slower growing cells as a result. This directional growth can occur via a plant's response to a particular stimulus, such as light (phototropism), gravity (gravitropism), water, (hydrotropism), and physical contact (thigmotropism).

Plant growth and development are mediated by specific plant hormones and plant growth regulators (PGRs). Endogenous hormone levels are influenced by plant age, cold hardiness, dormancy, and other metabolic conditions; photoperiod, drought, temperature, and other external environmental conditions; and exogenous sources of PGRs, e.g., externally applied and of rhizospheric origin.

Morphological Variation

Plants exhibit natural variation in their form and structure. While all organisms vary from individual to individual, plants exhibit an additional type of variation. Within a single individual, parts are repeated which may differ in form and structure from other similar parts. This variation is most easily seen in the leaves of a plant, though other organs such as stems and flowers may show similar variation. There are three primary causes of this variation: positional effects, environmental effects, and juvenility.

Evolution of Plant Morphology

Transcription factors and transcriptional regulatory networks play key roles in plant morphogenesis and their evolution. During plant landing, many novel transcription factor families emerged and are preferentially wired into the networks of multicellular development, reproduction, and organ development, contributing to more complex morphogenesis of land plants.

Developmental Model Organisms

Much of developmental biology research in recent decades has focused on the use of a small number of model organisms. It has turned out that there is much conservation of developmental mechanisms across the animal kingdom. In early development different vertebrate species all use essentially the same inductive signals and the same genes encoding regional identity. Even invertebrates use a similar repertoire of signals and genes although the body parts formed are significantly different. Model organisms each have some particular experimental advantages which have enabled them to become popular among researchers. In one sense they are "models" for the whole animal kingdom, and in another sense they are "models" for human development, which is difficult to study directly for both ethical and practical reasons. Model organisms have been most useful for elucidating the broad nature of developmental mechanisms. The more detail is sought, the more they differ from each other and from humans.

1. Plants:

    - Thale cress (Arabidopsis thaliana)

2. Vertebrates:

    - Frog: Xenopus (X.laevis and tropicalis). Good embryo supply. Especially suitable for microsurgery.
    - Zebrafish: Danio rerio. Good embryo supply. Well-developed genetics.
    - Chicken: Gallus gallus. Early stages similar to mammal, but microsurgery easier. Low cost.
    - Mouse: Mus musculus. A mammal with well-developed genetics.

3. Invertebrates:

    - Fruit fly: Drosophila melanogaster. Good embryo supply. Well-developed genetics.
    - Nematode: *Caenorhabditis elegans*. Good embryo supply. Well-developed genetics. Low cost.

Also popular for some purposes have been sea urchins and ascidians. For studies of regeneration urodele amphibians such as the axolotl Ambystoma mexicanum are used, and also planarian worms such as Schmidtea mediterranea. Organoids have also been demonstrated as an efficient model for development. Plant development has focused on the thale cress Arabidopsis thaliana as a model organism.

Importance of Developmental Biology

Developmental biology (DB) is an expansion of traditional embryology, which studies the development of a single-cell fertilised egg to a multicellular free-living organism. Today, DB research covers beyond embryogenesis to include growth, sexual maturation and ageing.

DB affects our life in many aspects.

One of the key issue related to our life and the sustention of species is the control of fertility and infertility. The understanding of oestrous cycle and the development of sexual organs and gametes helps us control conception and increase fertility. Birth control pills and condoms are both effective tools for population control. The development of in vitro fertilisation (IVF, i.e. test-tube baby) technology is another good example to show how DB research can change our lives. The desire of having their own baby was often a dream that never came true for infertile couples.

But since the development of IVF and other assisted reproductive technologies, many couples regain their hope to have biological babies. The success of IVF in mammals, including human, was based on the basic understanding of fertilisation mechanisms of marine invertebrates like sea urchins. From those studies, we appreciate the importance and regulation of ionic flows during fertilisation. By manipulating those factors along with others, scientists gradually broke through the IVF barriers in mice, human and other larger mammals. IVF is thus one of the most fascinating stories of how DB research can improve the wellbeing of human life itself.

How could a single-cell fertilised egg develop into a multicellular functional adult is a key fundamental question asked by developmental biologists. Throughout centuries of research, we understand that cells in earlystage embryos have the full potential to become all types of adult cells that are called “totipotency”. The totipotency is gradually lost in later stage embryos, but they still retain “pluripotency” to become multiple types of cells.

Eventually, embryonic cells limit their developmental potential to one specific type of cells. However, we know that even terminal differentiated somatic cells contain the same genome as that of a fertilised egg.

Theoretically, it can be totipotent if properly reprogrammed back to the original state. Although the astounding generation of Dolly Sheep proves the principle by somatic nuclear transplantation experiments, low success rate and unhealthiness of cloned animals imply that reprogramming the differentiated genome to its default state is not reached by using this method.

Many animals, like planarians, salamanders and zebrafish can replace missing body parts. In contrast, mammals only have a limited regenerative ability in tissues and organ-like skin epithelial layers, hair and liver. This may lead to severe health problems like fibrosis, accumulation of extracellular matrix proteins like collagens in heart and other organs. Heart injuries often lead to scars and subsequent fibrosis. Fibrosis results in the stiffness of heart, impaired function and eventually heart failure.

The studies of fibrosis prevention and resolution have attracted the attention of medical professionals, as well as DB researchers. DB researchers often use animal models capable of regeneration to understand how the regeneration is induced and accomplished.

They are now more eager to apply the gained knowledge to screen drugs or devise tools or strategies to repair injure tissues or organs in mammals and human that has been established in a new discipline called “regenerative medicine.”

During development, a group of cells called stem cells not only can become other types of cells, but they can renew themselves to keep their stemness. The most competent stem cells are embryonic stem cells (ESCs). ESCs are cells from the inner cell mass of a blastocyst. They contain pluripotency that allows them to be induced into many other types of cells in vivo or in vitro and is potentially a great cell source to be cultured to tissues, or even organs to be used in regenerative medicine.

Developmental biology (DB) is an expansion of traditional embryology, which studies the development of a single-cell fertilised egg to a multicellular free-living organism. Nowadays, the DB research covers beyond embryogenesis to include growth, sexual maturation and ageing.

However, it is unavoidable to destroy a blastocyst while collecting ESCs. It becomes a major hurdle that prevents the practical use of ESCs due to ethical and religious reasons. Terminal differentiated somatic cells still retain the same genome as that of fertilised eggs. Triumphantly, we now can reprogram somatic cells like fibroblast cells to become stem cells called inducible pluripotent stem cells (IPSCs). Using IPSCs different types of cells, including blood, neuronal cells, muscle cells, germ cells and others have been created in vitro.

The next challenge will be how to consistently produce those cells to be reliably used in regenerative medicine. IPSCs may also push the development of medicine to the next level. Traditionally, the development of drugs relies heavily on the use of immotile cell lines and animal models, mainly mice. However, only less than 1% of drugs tested in clinical trials are approved. The failure of drug testing can be for a variety of reasons.

The suitability of screening model is certainly one of the issues. Although mice resemble human in many ways, they are not human. Even in humans, there are variations between different people.

In addition, drugs entered the market are suitable for most people that may not fit you well. Therefore, the idea of "precision medicine" emerges. The IPSCs may fit in the precision medicine nicely. We can use patient derived IPSCs for drug testing before administrating to patients. We can also use human-derived IPSCs for drug screening as well. However, we still have a long way to go and proper animal models are still needed.

One of the most exciting discoveries in biology lately is the invention of CRISPR/Cas9 technology, which allows much more convenient targeted genome editing in cells and organism levels. DB researchers quickly apply this technology to edit genomes of different model systems, as well as for IPSCs. It is then of great potential to correct human genetic diseases with known mutation(s) using CRISPR/Cas9 if it is proven to be safe and effective in the future.

Overall, although most DB research still focuses on the basic regulations of developmental processes, developmental biologists are prompt to apply their findings or practical use in both medicine and agriculture.

DB asks Profound Questions at the Level of whole Organisms or Organs

DB investigates questions such as "how does the kidney or brain develop?", or "how do limbs or leaves achieve their characteristic shapes and positions?" To address such questions, a typical DB research strategy may start by identifying the genes or gene networks regulating the respective developmental processes in a chosen animal or plant. These genes can then be functionally manipulated or eliminated to study the resulting developmental aberrations. The findings often allow deductions about how the involved genes and processes function in health; they may also reveal parallels to clinical cases of human developmental disorders, thus directing further biomedical research into such conditions.

To investigate processes from the genetic level all the way up to the organism/organ level, DB must be highly inclusive and interdisciplinary, making active use not only of genetics but also biophysics, biochemistry, cell biology, physiology and anatomy. In this way, it drives discoveries at the various levels of complexity, acts as an umbrella discipline that can provide a common focus towards essential biological questions and builds bridges to clinicians or plant/animal breeders who tend to think at the organism/organ level.

DB makes Strategic uses of Model Organisms

Most DB research does not use human embryos but covers the breadth of the animal and plant kingdoms. This ambition might seem to bear the risk of overstretching our research capacities, but it is, in fact, a great strength of DB and gold mine for discovery. It turned out that many genes and functional gene networks that steer fundamental biological processes have ancient evolutionary origins and are still being used by very different species for similar purposes; ~75% of human disease genes have a counterpart in fruit flies and ~50% of yeast genes can be functionally replaced with human genes.

Capitalising on this principle of 'deep homology', highly efficient and cost-effective, hence economically responsible research can be done in smaller organisms, such as worms, flies or even yeast. The genes and concepts learned can then be tested in mammals (most frequently mice) and eventually used for clinical trials. This discovery pipeline has led to significant understanding of human biology and disease, as evidenced by an impressive number of Nobel Prizes in Physiology and Medicine awarded to scientists working with these "model systems".

Work Done by DB so far for us

DB research starts with the fertilisation of egg cells; studying the underlying processes has provided the foundations for much of what fertility clinics can do these days. DB investigates how fertilised egg cells divide in regulated manners to grow into full-size bodies, how the cells formed in this process communicate in meaningful ways to become different from each other, migrate, change shape and attach to each other, thus assembling into tissues and complex organs. Many of these processes are needed again during wound repair and DB research helps to speed up wound healing, prevent scars and overcome chronic wounds.

Also 'tissue engineering', which aims to grow replacement tissues in a plastic dish, is essentially guided by DB research. In cancer, cells lose their identity, divide excessively, detach from their local environments and migrate to form metastases. Much of this understanding that can instruct cures to contain these aberrant cells, comes from DB research. Tissues keep so-called stem cells which can be re-activated in orderly manners to divide and grow replacement tissues. There are high hopes from stem cell research, for example, to replace cartilage in arthritis or damaged discs, or brain cells in dementia, much of which is guided by the vast knowledge gained through DB.

The applications of DB go far beyond biomedical research. For example, understanding plant development provides a means to speed up breeding processes, such as optimising root systems, plant size or flowering time, thus contributing to the efforts of achieving sustainable food security in times of over-population. Furthermore, understanding environmental influences on development, such as temperature-dependent sex determination in turtles, has enormous importance for conservation biology, especially in times of increasing pollution and global warming.

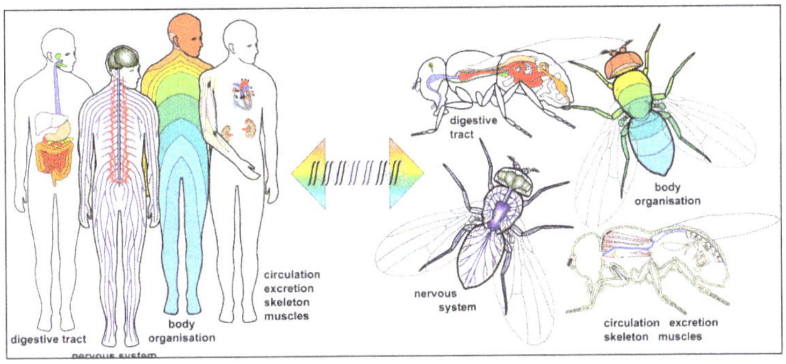

In conclusion, DB may appear as a mere academic discipline, but its value for society is enormous. This should make us think about a carefully balanced system of science funding. Current trends seem to favour clinical or industrial research performed to translate biological knowledge into an economic or societal benefit. But we must not overlook that fundamental research, such as in the field of DB, lays the long-term foundations for such developments.

The growth of biological endeavour has been strongly dependent of the kind of questions which were asked. The first basic and necessary approach is "what". The answers are mainly descriptive: what is the anatomy of a frog? What are the species existing in the Hawaiian archipelago? Taxonomy is basically a descriptive discipline, but even the mere inventory of extant species on earth is far from complete and requires further investigations.

A second step in biological knowledge is "how". How does this work? How does this develop? That is the major, central approach in modern experimental biology. It is applied to all levels of organisation from molecules to ecosystems, and tries to get answers to questions like: how a gene controls a protein synthesis; how two molecules interact; how the germ layers of an embryo arise from a single cell; how so many Drosophila species coexist in the Hawaiian archipelago.

Experimental, functional analysis is at the core of modern biology, it implies an analytical approach and leads to the major discoveries. Any system investigated is subdivided into parts, which are easier to handle and to analyse. In this respect there is both a temptation and a danger of reductionism. Reductionism will try to explain the functioning of a complex biological system by the mere addition of its known parts, while an increase in complexity always leads to the appearance of new, emerging, unexpected properties. The sequencing of entire genomes is certainly a major progress in biological knowledge, but it will not explain directly the realisation of an organism. Indeed, sequencing is more a "what" question than a "how" one. Since biological processes are integrated at different levels of complexity (molecules, cells, organisms, populations and ecosystems) and since, at each level, new emergent properties appear, it is both necessary and justified to tackle simultaneously all these levels. There should not be any hierarchy among biological disciplines, but only discrimination between good innovative science and poor repetitive science.

The third kind of question is "why". Why are there bacteria and elephants? Why micro-organisms and Metazoa? Why so many (over 1000) Drosophila species in Hawaii? Why species at all? Why is sometimes considered a crazy question, but it is also a pervading interrogation since the beginning of historical times. Why are there living beings on earth? Why are there humans? In the absence of scientific knowledge, philosophers tried to answer these questions and the recurrent explanation in almost all civilisations was creation: there was nothing to understand further. Only the accumulation of what and how answers during the last three centuries has permitted the birth and the growth of another explanation: all living organisms share a common ancestry and have evolved and diversified during almost four billion years of earth history. Biological evolution, not creation, is the general and now well recognised process. Much however still remains to be explained and this is the goal of a mature discipline in life sciences: evolutionary biology. It is a difficult discipline for at least two main reasons. First, as stated above, it must integrate, interpret and explain all kinds of biological information at all levels of integration. Second, being basically an historical approach, it is in many cases not amenable to an experimental analysis. In other words, many hypotheses and interpretations in evolutionary biology cannot be falsified in the sense of Popper. Evolutionists, using a comparative method, are proposing plausible interpretations, often implementing mathematical techniques. These models may be falsified or improved by further observations.

Being based on a principle of inheritance with change over generations in natural populations, evolutionary biology needed to integrate all the information from genetics, especially from population genetics. Finding now the same genes in different species and distant taxa has been a further "verification" of the evolutionary theory and has permitted to work out molecular phylogenies, providing irreplaceable historical information. More recently major conceptual progresses have been made at a higher integrated level, in interpreting the evolution of life histories. This rapidly expanding field, at the crossroads between observations and theories, tries to explain why bacteria and elephants coexist in nature.

During the maturation years of evolutionary theory, embryological information has played a fairly marginal role, in spite of the fact that most extant species are multicellular. Having realised that identical genes are expressed in very different groups and lineages, developmental biologists are now aware that they must integrate evolutionary information in their research, at least the historical and phylogenetic aspects. Evolutionary biologists, on the other hand, will discover the diversity and richness of problems raised by the development of highly complex multicellular organisms. Time is coming for a convergent evolution of the two disciplines.

The Evolutionary Theory at the Beginning

With the exception of a few creationists, biological evolution is now recognised as a well-established occurrence. Such was not always the case. Some ideas about evolution may be found among ancient philosophers as well as in the works of Buffon but the first clear formalisation dates back to Lamarck. Following Linné, the classification of plants and animals used morphological similarity in order to establish hierarchical taxa at various levels (e.g., genera, families, orders). The general interpretation was similarity by kind, that is, those animals looking alike, such as horses and donkeys, had been created to perform similar functions in nature. The basic idea of Lamarck was that similarity was more easily explained by a common ancestral origin. Of course, such an hypothesis, called transformism, needed a mechanism. At that time, nothing was known about heredity. It was however quite obvious that different environments and ways of life might result in large variations of phenotypes (what we call now phenotypic plasticity). Lamarck proposed that such changes, induced in parents, would be passed on to offspring, a mechanism called inheritance of acquired characters.

Sometime later, Geoffroy St. Hilaire, comparing different animal phyla, tried to establish a single plan of organisation. His comparison of the anatomy of a vertebrate and of an arthropod (a lobster) remains a well-remembered example. His idea was to find homologous organs in distant phyla, an argument for a single plan of organisation, and of course for a possible evolution from a single ancestor. This "théorie des analogues" (homologous organs were called analogous) was fiercely fighted by Cuvier, a bright comparative anatomist, but a fixist and a creationist.

The idea of Evolution was extended three decades later by Darwin. Again, nothing was precisely known about heredity, and Darwin believed in the heredity of acquired characters. His theory of gemmules, elaborated later, also turned out to be wrong. The merit of Darwin, however, was to suggest and argue for a mechanism of evolution, which is natural selection. Still now natural selection, with its multiple aspects, appears as the central mechanism of biological evolution.

The Evolutionary Tree of Life

In modern science, a broad diversity of information helps to feature how life was born on earth, and how and when the different groups now observed in nature originated. Such inferences and reconstruction are mainly based on two very different kinds of data: the fossil record and molecular phylogenies.

Broadly speaking, the surface of our planet solidified about four billion years ago. In a relatively short period (500 million years) prebiotic systems evolved and produced cells surrounded by a membrane, more or less similar to extant bacteria. Typical bacterial constructions, the stromatolithes, are now observed along the seashore, but already existed three billion years ago. The further evolution of Prokaryotes (Bacteria and Archea) was fairly slow with little phenotypic changes. A major event, which took place about two billion years ago, was the emergence of Eukaryotes, that is, of cells with a nucleus and many other cellular characteristics. The origin of Eukaryotes is still debated. Molecular investigations have shown that, according to the kind of genes which are compared, Eukaryotes appear to be closer either to Bacteria or to Archea. Of course the fossil documents are very scarce. A novel theory now spreading among biologists, assumes that the first ancestral Eukaryote arose from a symbiosis between a Bacterium and an Archea: the membrane of the internal symbiont produced the nuclear membrane, and the two genomes progressively fused into a single one, within the nucleus. The acquisition of a bigger and more complex genome was followed by further major changes. We may recall the origin of mitochondria from another cytoplasmic bacterial symbiont and the acquisition of photosynthesis (chloroplasts) from symbiotic cyanobacteria.

For a long evolutionary time, Eukarya remained unicellular but produced a broad diversity of phyla still found in the extant world. These unicellular eukaryotes are often classified in a single cluster, or kingdom, called Protista, although they include a diversity of clades and are obviously paraphyletic. During their evolution, Eukaryotes acquired progressively new properties and organelles, such as chromosomes, mitosis and presumably meiosis. Possible mechanisms can be suggested from a comparison of what now exists in the living world while the fossil record is useless.

Multicellularity - A Major Recurrent Evolutionary Change

Most species in the extant world are multicellular organisms (plants and animals) living on continents. About a billion years ago, all organisms were unicellular and lived in oceans. The progressive colonisation of terrestrial biota is quite well documented thanks to fossils, comparative anatomy and phylogenetic reconstruction. This was accompanied by spectacular new phenomena such as increase in body size, diversification of life histories and ecological niches, and more complex interactions between species.

Multicellularity is a recurrent phenomenon which occurred independently several times. Multicellular animals (i.e. Metazoa) appear as a monophyletic clade. Terrestrial plants are another one, and also Fungi. Multicellularity is also observed in algae, among Protista (e.g. slime molds) or among Bacteria (stromatolithes).

Multicellularity raises a major challenge for evolutionary theory. In terms of Darwinian fitness, multicellularity implies, in a primitive stage, that some cells will not contribute to the next generation: they remain somatic for the benefit of their sisters. In other words, they become altruistic and the evolution of altruism, as observed in animal societies, was already a challenge for Darwin. Another problem is that multicellularity is accompanied by a delay in reproduction, and a decrease in the number of offspring, both resulting in an apparent decrease in the fitness of the population.

Explaining the Origin of Multicellularity

The paradox of altruistic behaviour in animal societies was explained by the concepts of kin selection and inclusive fitness. An altruistic gene may invade a population if the sacrifice of one female (males are not considered in demographic analyses) permits the survival of more than two sisters (two sisters will harbour on average one altruistic allele, since the genetic similarity among sisters is 50%). In the case of a multicellular organism, altruism will be easier since all cells share the same genotype. However, the reduction in fitness of a multicellular variant will be more pronounced since, in most cases, many cells fail (renounce) to reproduce.

For example, among Volvocales (phytoflagellates) different related species exist, some remaining unicellular while others build multicellular entities. Under favourable laboratory conditions, a unicellular form will rapidly eliminate a multicellular one, thanks to a higher proliferative capacity accompanied by a low mortality. The picture changes completely if some predator is introduced. Big individuals will be protected by their size while unicellular ones will be eaten and will decrease in number. This simple example illustrates how different species may coexist. It is not possible for a single species to acquire all possible advantages without paying for some inconveniences; otherwise only one species would ultimately survive on earth.

A rapid rate of reproduction favouring the invasive capacity is accompanied by a vulnerability to predation. A big size, on the other hand, may protect from predation and may offer some other advantages, but will imply a longer generation time and a lesser rate of proliferation. More generally any adaptive change in one direction which provides some benefit in terms of fitness, is likely to be accompanied by a cost from another point of view. Evolutionary biologists describe such negative correlations as trade-offs. During the last two decades, major conceptual progresses have been made using the concept of trade-off, for interpreting the diversity of life histories. In the extant world, trade-off is the ultimate cause of species coexistence, and it explains why it is possible to be a weed or a tree, a mouse or an elephant. Bacteria have remained very small and quite simple since they appeared more than three billion years ago, but they are still doing very well. Metazoa are much more complex and much bigger. But as stated by all evolutionists, complexity is not an indication of a general tendency toward some progress. The human species, which may be considered as a summum of complexity is, in other respects, very vulnerable. Multicellularity has certainly been a means for producing a bigger size, thus avoiding predation or becoming a predator. But other interpretations have been considered, for example, a better metabolic utilisation of the environmental conditions.

Another problem with multicellularity is that it implies some change at the genetic level. A modification in the genetic program is likely when two cells which arise from a mitosis, remain together and eventually change their phenotypes and their functions. Such changes however do not modify the genome in a permanent way, like mutations. More likely they involve changes in the way genes are regulated by epigenetic processes. Cell differentiation, which is a major but still poorly understood issue in developmental biology, is almost exclusively an epigenetic, more or less reversible phenomenon. Theoretical evolutionists now try to understand how such epigenetic changes (sometimes called epimutations) may spread in a population. Much remains to be done in this way of thinking.

Analogies and Homologies

An ancient but still valid and modern debate in evolutionary biology, and also for understanding development, is to distinguish analogies from homologies.

Convergent, recurrent adaptations are likely to occur when a strong selective pressure exists: this will produce analogous organs or functions. For example the wing of a fly and the wing of a bird share a common function, bear the same name but have completely independent origins. The problem becomes however more complex if we compare bats and birds, two kinds flying vertebrates. In both cases, the forelimb has been transformed into a wing. In this respect, wings are homologous to the anterior legs of Amphibia and Reptilia, and can be thus considered, at the forelimb level, as homologous. But the flight function results from convergent, independent adaptations that are the transformation of an anterior leg into a wing: in this respect, the wings of birds and bats are analogous organs. The above example is easy to understand, but in other cases, the origin of an organ by evolutionary transformation may be very difficult to solve. For example, the wings of insects are considered as derivatives either of segmental gills or of lateral expansions of thoracic tergites. It is however generally assumed that all insect wings share a common origin, i.e. are homologous.

The development of molecular biology and the search for molecular phylogenies has brought a new interest concerning the problem of homologous or analogous genes and functions. By comparing the sequences (DNA or amino-acids) of two genes, it is generally easy to decide whether they are homologous, deriving from a common ancestor or not. An identity of function, for example the same EC code for an enzyme, does not however implies homology. For example, all alcohol dehydrogenases (ADH) share the same code: EC 1.1.1.1. However, at the molecular level, the ADHs of yeast, maize and mammals are homologous while the Drosophila ADH is a much smaller, analogous molecule.

Evolutionists are also aware that, in the course of Evolution, gene duplications have been common providing an opportunity for acquiring new functions. This has led to a classical problem in phylogenetic analyses. When comparing genes in different species, there are two kinds of homologies: orthologous genes derive directly from a single ancestor and have diverged after the last specific split; paralogous genes have different ancestries in closely related species, since they originated from much older gene duplication. It is misleading to use and compare paralogous genes while believing they are orthologous.

The Origin and Evolution of Developmental Novelties

Evolutionary biologists are aware that adaptations are constrained by historical features, especially by phylogenies, so that evolutionary changes are more akin to an erratic tinkering than a general directional trend. From a functional point of view, we easily understand a loss of function, for example the origin of snakes from lizard-like ancestors. We also understand the transformation of an anterior leg into a wing, especially because we recognise the same bones in both organs. It is far more difficult to imagine how a tetrapod vertebrate might evolve into a hexapod. Most scientists will consider such a change as impossible. In other words, once the tetrapod state was acquired in terrestrial vertebrates, it became a strong developmental constraint which did not permit eventual major changes.

The observation of such constraints has led many authors to say that Darwinian theory explained microevolution (that is small adaptive variations) but not macroevolution, defined as the appearance of major novelties, such as limbs or head. This opposition still remains a debate among evolutionists, but it is progressively solved by a better understanding of both phylogenies and comparative developmental genetics.

In the Chordate phylum, it is likely that an ancestral state was like the extant Amphioxus, deprived of limbs. Then limbs appeared presumably as lateral expansions along the body. Why only two pairs and not three have been retained in the derived groups is not understood.

We have more information concerning the origin of the head in Vertebrates. Again, the Amphioxus-like ancestor was presumably deprived of a differentiated head, without eyes or mouth and with only a slight anterior dilatation of the neural tube. In the course of evolution, a major developmental novelty appeared, that is, a special derivative of the nervous tube, the neural crest. Modern developmental investigations have shown that the neural crest plays a major role in the organisation of the nervous system in Vertebrates and especially in the building of the head, including the brain and the skull. Without the neural crest, the evolution of Vertebrates toward a bigger size and the acquisition of a complex brain would have been impossible. How the neural crest itself originated during chordates evolution is not known. Future progresses of developmental genetics should shed some light upon this major problem.

The Antero-Posterior Axis: A Major Success of Developmental Genetics

Evolution and development (Evo-Devo) is becoming a fashionable preoccupation in modern biology. This is due to the fact that evolutionists discover development, while developmental biologists discover evolution. I believe that a major reason for this common interest originated in the unexpected discovery that the early stages of embryonic development and especially the establishment of the antero-posterior axis were homologous features in Drosophila and Vertebrates. More precisely, Drosophila geneticists progressively identified a series of genes (called homeotic) which were responsible for the fate of successive body segments and for their differentiation into head, thorax and abdomen. Surprisingly, there was a colinearity between the order of these homeotic genes on the third chromosome, and of their expression along the antero-posterior axis of the embryo. When these genes were cloned, it was rapidly found that they also existed in Vertebrates, and surprisingly in the same order on the chromosome. More precisely these genes, called Hox genes, form four paralogous clusters on four different chromosomes. The existence of these four clusters confirms the hypothesis put forward three decades ago that, during Vertebrate evolution, two successive tetraploidisations took place.

The similar order of the genes in Vertebrates and Drosophila (and in several other phyla) implies for example that the same genes are expressed in the head of both taxa. It was tempting at the beginning to assume that the head of a fruit fly and the head of a mouse might be homologous organs. However this conclusion would be wrong. Drosophila and Vertebrates are distantly related Metazoa. Vertebrates belong to chordates within Deuterostomia while Drosophila belongs to Ecdysozoa within Protostomia. We have seen that in chordates, an ancestral state was very likely deprived of a differentiated head, and that the Vertebrate head was a new acquisition depending on the neural crest. Besides Arthropods, Ecdysozoa comprise other phyla which, like nematodes, do not possess a differentiated head. The absence of a head is likely a primitive state; in other words, the insect head is also likely to be a new acquisition, rather than being homologous to the vertebrate head.

The fact that the same genes are expressed during head development in both Drosophila and Vertebrates must be considered as an analogy and an evolutionary convergence. Why this convergence used the same genes to produce analogous organs is unclear. One interpretation is that evolution was constrained by a limited set of developmental genes and was obliged to reutilize the same genes for building new organs and new functions. It is also possible that the functional order of the homeotic genes along the chromosome was an additional constraint, preventing the building of a head with completely unrelated genes.

Another Developmental and Evolutionary Paradigm: The Origin of the Eyes

Eyes are obviously made to see, and they are found in most animal phyla. In some cases, they are very complex organs, performing their function close to perfection. In the early time of Darwinism, eyes were often taken as an argument to say that such complex structures could not proceed from a gradualistic evolutionary process. In other words, and in the same way that finding a watch implied a watchmaker, the observation of the perfect eye of a bird implied some intention in evolution, and has long been a creationist argument. Improvements in scientific knowledge have progressively changed the way we see and understand the evolution and development of the eye.

It was recently shown by computer simulation that a simple flat photoreceptor could evolve progressively into a complex spherical eye with a lens, by fixation of random mutations through natural selection.

The diversity of eyes which are found for example among molluscs is best explained by a gradualistic evolution and different levels of selective pressure. The similarity of complex eyes as found in an octopus and a vertebrate is simply due to convergent adaptive evolution. If eyes are recurrent adaptations occurring in diverse phyla, the question is: Does Developmental Biology provide some interesting insight about eye evolution.

A positive answer was provided when it was shown in Drosophila that an over expression of a gene (the protein of eyeless) could induce the formation of ectopic eyes. The eyeless gene is homologous to Pax 6 in vertebrates, and mutants of the two genes result in eye abnormalities in fruit flies and vertebrates respectively. A possible but wrong conclusion would be that, despite their anatomical differences, eyes in insects and vertebrate are homologous. What is homologous in the two phyla is the basic process of inducing some cells into the photoreception pathways, with the production of the photosensitive protein, opsin. During the development of the eyes, the two genes are used again for morphological differentiation, including lens production in vertebrates, but these later functions are pure analogies.

How various genes, having a diversity of basic functions, may be recruited to perform a new task is best illustrated by the lens proteins (crystalline) in vertebrates.

The new function just consists in making a transparent solution within the lens, and many proteins can do that. Lens proteins are very diverse: some are constitutive, i.e., general proteins expressed in all species. But others are metabolic enzymes which have found a new site of expression, but no metabolic function: for example, lactate dehydrogenase in birds, alcohol dehydrogenase in camels, aldehyde dehydrogenase in macroscelids.

Sexual Reproduction: A Poorly Understood and Recurrent Evolutionary Dilemma

Already for Darwin, sexual reproduction and sexual selection appeared as a specific, but difficult problem. Since these times much attention has been paid by both evolutionists and developmental biologists to the understanding of sexuality, but a deep gap still persists between the two approaches.

In most animal species sexes are separated, about 50% of individuals are males, 50% are females. From an ecological and evolutionary point of view, this proportion is not an optimum. Optimising selection should favour species in which there are more females than males. Why this is not so is generally explained by genetical constraints. In most species, sex appears to be transmitted as a simple Mendelian character, leading to a 50-50 segregation. Only in a few groups, such as reptiles or haplodiploid arthropods, sex is not genetically determined, and the sex ratio may vary, often in an adaptive way, i.e. producing more females than males.

Indeed, the occurrence of sexual reproduction is, within evolutionary theory, a puzzle and a dilemma. It is commonly argued that sexual reproduction which implies not only fertilisation, zygote production and embryonic development, but also genetic exchanges among different individuals of the same species, is a means for conserving the genetic diversity, keeping the genetic load at an acceptable level and permitting eventually new adaptations and evolutions.

But all these are supposed long-term advantages, in opposition with short-term disadvantages. Although most animals are gonochoric (two sexes) many are parthenogenetic, that is reproduction occurs without fertilisation and by a direct development from female gametes. Parthenogenesis is a recurrent phenomenon which has evolved independently from bisexual species in many groups.

From a theoretical point of view, it is easy to show that a parthenogenetic mutant, appearing in a bisexual species, should invade the population and outcompete the gonochoric form. In other words, parthenogenesis should be the general way of reproduction, and why it is not so remains a dilemma. Evolutionists are obliged to suggest ad hoc arguments, and the most general one is that parthenogenetic organisms are made of clones with a low genetic diversity. This should prevent their adaptation to changing environments and result in high rates of extinction. In other words, the short-term advantage of parthenogenesis should be counterbalanced by a long-term disadvantage. It is also possible, for example in Mammals, that parthenogenesis is not possible due to genetic imprinting and the differential expression of genes coming from the sperm or the oocyte in the embryo.

Developmental biologists have also long been fascinated by sexuality. In Vertebrates, we know than sex hormones play a major role in the differentiation of primary and secondary sexual characters. In Drosophila, on the other hand, each cell is autonomous and capable of expressing independently its own sexual genotype and no hormones are involved.

Sex differentiation has been analysed not only by classical experiments of developmental biology such as grafts and extirpations, but also by genetic means. Indeed, it has been the first great success of developmental genetics. To summarise the results shortly, sex is determined in Drosophila by a succession of genetic cascades acting within each cell. In Mammals, the primary induction of the male hormone in the testis will eventually determine the whole male phenotype. In the female, the absence of male hormone permits the differentiation of the female phenotype.

Sexes are phenotypes which are very strongly constrained, or canalised, by natural selection. Individuals must be normal males or normal females for mating and reproduction. Genetic dissections of sexual phenotypes have produced intermediate individuals which are almost always abnormal and sterile. So the question is: 'How is the genetic determinism of sex able to evolve.'

The strong constraints existing on phenotypes suggest that, as for the antero-posterior axis of early embryos, we should find genetic homologies in sex determination of distantly related taxa. Developmental genetics did show exactly the reverse. The genetics of sex determination is a fast evolving trait; sexual phenotypes are pure analogies. Different evolutionary pathways are visible not only between distant taxa (e.g. arthropods and vertebrates) but also within arthropods: for example, a sex hormone, the androgenic hormone is responsible of male differentiation in Crustacea. Even within the same insect order, major differences are found: genetic determinisms of sex in houseflies or mosquitoes are very different from that of Drosophila.

Here is a clash between developmental and evolutionary biology. Developmental genetics has revealed a high diversity among mechanisms of sex determination, presenting sex as a relatively fast evolving trait producing analogous phenotypes. Evolutionary biologists, however, cannot understand and explain how it is possible to evolve from one genetic system to another one, since any genetic change is likely to produce abnormal, sterile individuals. The only possibility we can see for the moment is that such a transition might occur by the intermediary of a relatively neutral form, that is, a parthenogenetic lineage. In such a lineage, stabilising selection on sexual phenotypes would be released, permitting some genetic changes. In the long-term, but for unknown reasons, a new male phenotype could appear with an original genetic basis. Of course, this hypothesis requires further theoretical analyses, in order to understand why and how a neo-male might occur.

Early Developmental Processes

Cleavage

After fertilization, the development of a multicellular organism proceeds by a process called cleavage, a series of mitotic divisions whereby the enormous volume of egg cytoplasm is divided into numerous smaller, nucleated cells. These cleavage-stage cells are called blastomeres. In most species (mammals being the chief exception), the rate of cell division and the placement of the blastomeres with respect to one another is completely under the control of the proteins and mRNAs stored in the oocyte by the mother. The zygotic genome, transmitted by mitosis to all the new cells, does not function in early-cleavage embryos. Few, if any, mRNAs are made until relatively late in cleavage, and the embryo can divide properly even when chemicals are used experimentally to inhibit transcription. During cleavage, however, cytoplasmic volume does not increase. Rather, the enormous volume of zygote cytoplasm is divided into increasingly smaller cells. First the egg is divided in half, then quarters, then eighths, and so forth. This division of egg cytoplasm without increasing its volume is accomplished by abolishing the growth period between cell divisions (that is, the G_1 and G_2 phases of the cell cycle). Meanwhile, the cleavage of nuclei occurs at a rapid rate never seen again (not even in tumor cells). A frog egg, for example, can divide into 37,000 cells in just 43 hours. Mitosis in cleavage-stage Drosophila embryos occurs every 10 minutes for over 2 hours and in just 12 hours forms some 50,000 cells. This dramatic increase in cell number can be appreciated by comparing cleavage with other stages of development. Figure shows the logarithm of cell number in a frog embryo plotted against the time of development. It illustrates a sharp discontinuity between cleavage and gastrulation.

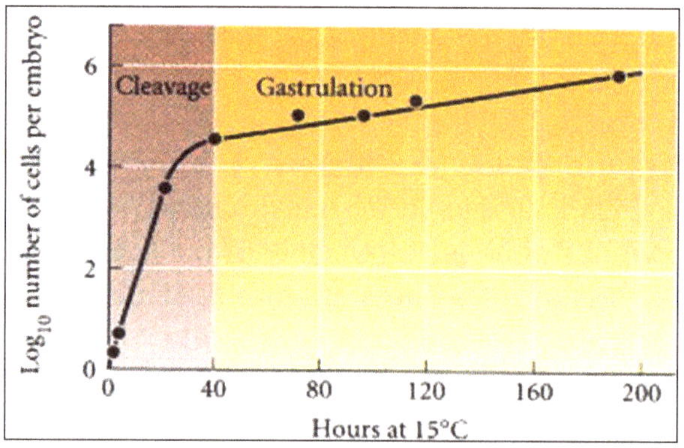

Rate of formation of new cells during the early development of the frog Rana pipiens.

One consequence of this rapid cell division is that the ratio of cytoplasmic to nuclear volume gets increasingly smaller as cleavage progresses. In many types of embryos (such as those of Xenopus and Drosophila, but not those of C. elegans or mammals), this decrease in the cytoplasmic to nuclear volume ratio is crucial in timing the activation of certain genes. For example, in the frog Xenopus laevis, transcription of new messages is not activated until after 12 divisions. At that time, the rate of cleavage decreases, the blastomeres become motile, and nuclear genes begin to be transcribed. This stage is called the mid-blastula transition. It is thought that some factor in the egg is being titrated by the newly made chromatin, because the time of this transition can be changed by experimentally altering the ratio of chromatin to cytoplasm in the cell. Thus, cleavage begins soon after fertilization and ends shortly after the stage when the embryo achieves a new balance between nucleus and cytoplasm.

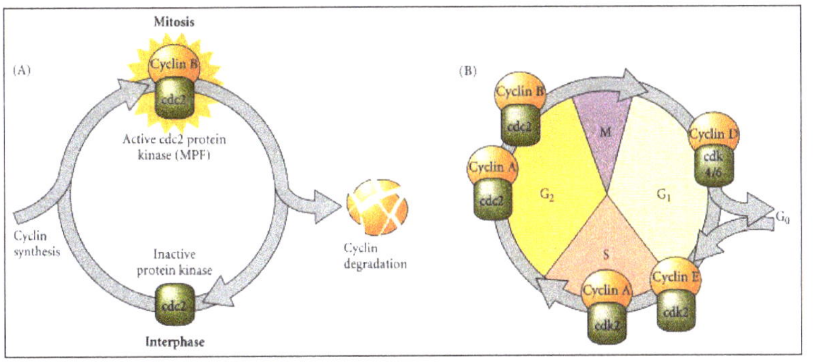

In figure, Cell cycles of somatic cells and early blastomeres. (A) The simple biphasic cell cycle of the early amphibian blastomeres has only two states, S and M. Cyclin synthesis allows progression to M (mitosis), while cyclin degradation allows cells to pass into S (synthesis) phase. (B) Cell cycle of a typical somatic cell. Mitosis (M) is followed by an "interphase"stage. This latter period is subdivided into G_1, S (synthesis), and G_2 phases. Cells that are differentiating are usually taken "out" of the cell cycle and are in an extended G_1 phase called G_0. The cyclins and their respective kinases responsible for the progression through the cell cycle are shown at their point of cell cycle regulation.

What causes this cyclic activity of MPF? Mitosis-promoting factor contains two subunits. The large subunit is called cyclin B. It is this component that shows a periodic behavior, accumulating during S and then being degraded after the cells have reached M. Cyclin B is often encoded by mRNAs stored in the oocyte cytoplasm, and if the translation of this message is specifically inhibited, the cell will not enter mitosis. The presence of cyclin B depends upon its synthesis and its degradation. Cyclin B regulates the small subunit of MPF, the cyclin-dependent kinase. This kinase activates mitosis by phosphorylating several target proteins, including histones, the nuclear envelope lamin proteins, and the regulatory subunit of cytoplasmic myosin. This brings about chromatin condensation, nuclear envelope depolymerization, and the organization of the mitotic spindle.

Without cyclin, the cyclin-dependent kinase will not function. The presence of cyclin is controlled by several proteins that ensure its periodic synthesis and degradation. In most species studied, the regulators of cyclin (and thus, of MPF) are stored in the egg cytoplasm. Therefore, the cell cycle is independent of the nuclear genome for numerous cell divisions. These early divisions tend to be rapid and synchronous. However, as the cytoplasmic components are used up, the nucleus begins to synthesize them. The embryo now enters the mid-blastula transition, in which several new phenomena are added to the biphasic cell divisions of the embryo. First, the growth stages (G_1 and G_2) are added to the cell cycle, permitting the cells to grow. Before this time, the egg cytoplasm was being divided into smaller and smaller cells, but the total volume of the organism remained unchanged. Xenopus embryos add those phases to the cell cycle shortly after the twelfth cleavage. Drosophila adds G_2 during cycle 14 and G_1 during cycle 17. Second, the synchronicity of cell division is lost, as different cells synthesize different regulators of MPF. Third, new mRNAs are transcribed. Many of these messages encode proteins that will become necessary for gastrulation. If transcription is blocked, cell division will occur at normal rates and at normal times in many species, but the embryo will not be able to initiate gastrulation.

The Cytoskeletal Mechanisms of Mitosis

Cleavage is actually the result of two coordinated processes. The first of these cyclic processes is karyokinesis—the mitotic division of the nucleus. The mechanical agent of this division is the mitotic spindle, with its microtubules composed of tubulin (the same type of protein that makes up the sperm flagellum). The second process is cytokinesis—the division of the cell. The mechanical agent of cytokinesis is a contractile ring of microfilaments made of actin (the same type of protein that extends the egg microvilli and the sperm acrosomal process). Table presents a comparison of these agents of cell division. The relationship and coordination between these two systems during cleavage is depicted in figure, in which a sea urchin egg is shown undergoing first cleavage. The mitotic spindle and contractile ring are perpendicular to each other, and the spindle is internal to the contractile ring. The contractile ring creates a cleavage furrow, which eventually bisects the plane of mitosis, thereby creating two genetically equivalent blastomeres.

Table: Karyokinesis and cytokinesis.

Process	Mechanical agent	Major protein composition	Location	Major disruptive drug
Karyokinesis	Mitotic spindle	Tubulin microtubules	Central cytoplasm	Colchicine, nocodazole[A]
Cytokinesis	Contractile ring	Actin microfilaments	Cortical cytoplasm	Cytochalasin B

Where, a = Because colchicine has been found to independently inhibit several membrane functions, including osmoregulation and the transport of ions and nucleosides, nocodazole has become the major drug used to inhibit microtubule-mediated processes.

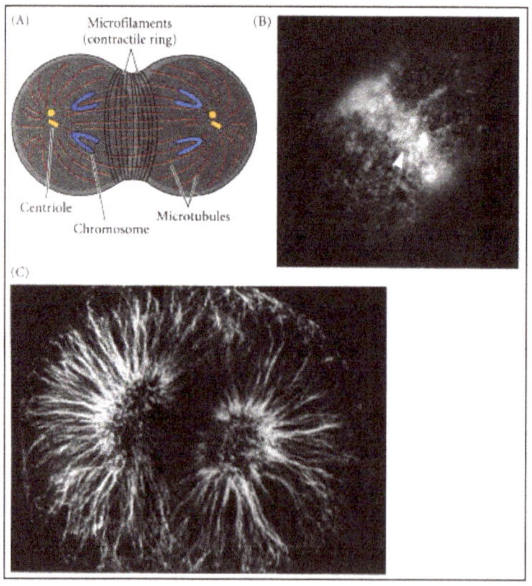

In figure, Role of microtubules and microfilaments in cell division. (A) Diagram of first-cleavage telophase. The chromosomes are being drawn to the centrioles by microtubules while the cytoplasm is pinched in by the contraction of microfilaments. (B) Localization of actin microfilaments in the cleavage furrow. Fluorescent labeling of the actin microfilaments shows the contractile ring in the first-cleavage furrow (arrowhead) of a telophase sea urchin egg. (C) Fluorescent labeling of tubulin shows the microtubular asters of a sea urchin egg during telophase of first cleavage.

The actin microfilaments are found in the cortex of the egg rather than in the central cytoplasm. Under the electron microscope, the ring of microfilaments can be seen forming a distinct cortical band 0.1 μm wide. This contractile ring exists only during cleavage and extends 8–10 μm into the center of the egg. It is responsible for exerting the force that splits the zygote into blastomeres; for if it is disrupted, cytokinesis stops. Schroeder has proposed a model of cleavage wherein the contractile ring splits the egg like an "intercellular purse-string," tightening about the egg as cleavage continues. This tightening of the microfilamentous ring creates the cleavage furrow. Microtubules are also seen near the cleavage furrow (in addition to their role in creating the mitotic spindles), since they are needed to bring membrane material to the site of membrane addition.

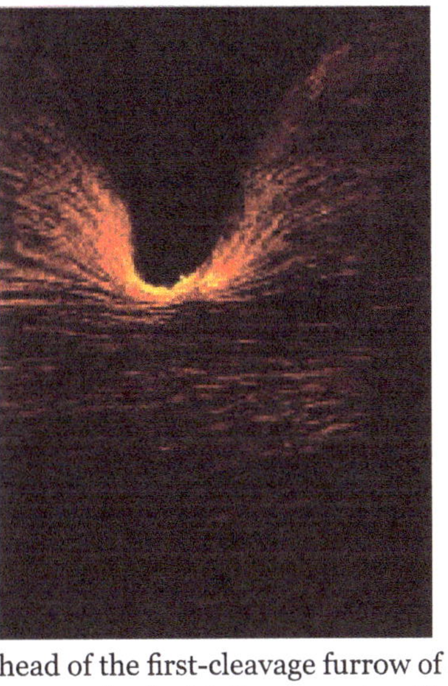

Microtubular array just ahead of the first-cleavage furrow of a dividing Xenopus egg. The microtubules are stained with fluorescent antibodies to tubulin.

Although karyokinesis and cytokinesis are usually coordinated, they are sometimes separated by natural or experimental conditions. In insect eggs, karyokinesis occurs several times before cytokinesis takes place. Another way to produce this state is to treat embryos with the drug cytochalasin B. This drug inhibits the formation and organization of microfilaments in the contractile ring, thereby stopping cleavage without stopping karyokinesis.

Patterns of Embryonic Cleavage

In 1923, embryologist E. B. Wilson reflected on how little we knew about cleavage: "To our limited intelligence, it would seem a simple task to divide a nucleus into equal parts. The cell, manifestly, entertains a very different opinion." Indeed, different organisms undergo cleavage in distinctly different ways. The pattern of embryonic cleavage particular to a species is determined by two major parameters: the amount and distribution of yolk protein within the cytoplasm, and factors in the egg cytoplasm that influence the angle of the mitotic spindle and the timing of its formation.

The amount and distribution of yolk determines where cleavage can occur and the relative size of the blastomeres. When one pole of the egg is relatively yolk-free, the cellular divisions occur there at a faster rate than at the opposite pole. The yolk-rich pole is referred to as the vegetal pole; the yolk concentration in the animal pole is relatively low. The zygote nucleus is frequently displaced toward the animal pole. In general, yolk inhibits cleavage. Figure provides a classification of cleavage types and shows the influence of yolk on cleavage symmetry and pattern.

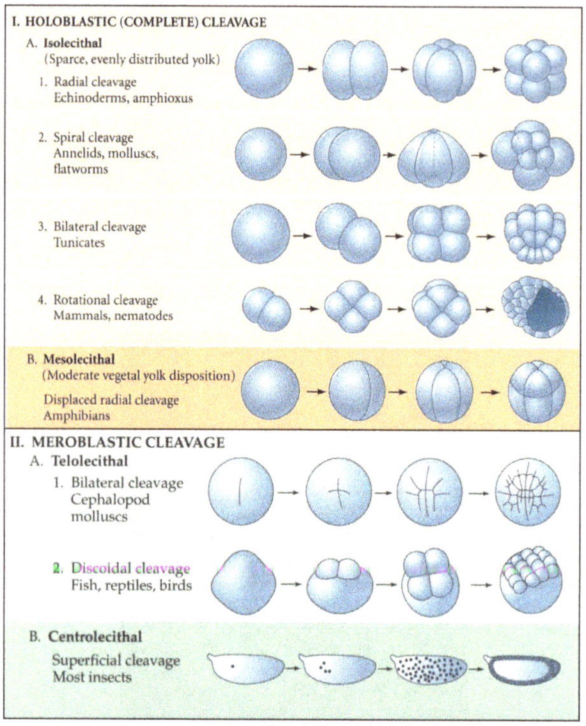

Summary of the main patterns of cleavage.

At one extreme are the eggs of sea urchins, mammals, and snails. These eggs have sparse, equally spaced yolk and are thus isolecithal. In these species, cleavage is holoblastic meaning that the cleavage furrow extends through the entire egg. These embryos must have some other way of obtaining food. Most will generate a voracious larval form, while mammals get their nutrition from the placenta.

At the other extreme are the eggs of insects, fishes, reptiles, and birds. Most of their cell volumes are made up of yolk. The yolk must be sufficient to nourish these animals. Zygotes containing large accumulations of yolk undergo meroblastic cleavage, wherein only a portion of the cytoplasm is cleaved. The cleavage furrow does not penetrate into the yolky portion of the cytoplasm. The eggs of insects have their yolk in the center (i.e., they are centrolecithal), and the divisions of the cytoplasm occur only in the rim of cytoplasm around the periphery of the cell (i.e., superficial cleavage). The eggs of birds and fishes have only one small area of the egg that is free of yolk (telolecithal eggs), and therefore, the cell divisions occur only in this small disc of cytoplasm, giving rise to the discoidal pattern of cleavage. These are general rules, however, and closely related species can evolve different patterns of cleavage in a different environment.

However, the yolk is just one factor influencing a species' pattern of cleavage. There are also inherited patterns of cell division that are superimposed upon the constraints of the yolk. This can readily be seen in isolecithal eggs, in which very little yolk is present. In the absence of a large concentration of yolk, four major cleavage types can be observed: radial holoblastic, spiral holoblastic, bilateral holoblastic, and rotational holoblastic cleavage. We will see examples of these cleavage patterns below when we take a more detailed look at the early development of four different invertebrate groups.

The absence of an incisive understanding of the action of toxicants on development has been in large part attributable to the absence of understanding of development itself. Until a few years ago, there was no understanding of a "developmental mechanism" at the molecular level although there were explanations at the cellular and tissue levels, such as "gastrulation is the mechanism by which the organization of the egg is transformed into the organization of the embryo." Recent advances in developmental biology have been substantial enough for scientists to be confident for the first time that some aspects of development in some organisms are understood at the molecular level. Protein components are identified, their functions in developmental processes are known, and the time and place in the embryo of expression of the genes encoding them are known. This knowledge greatly benefits elucidating the mechanisms of developmental toxicity.

The committee, in response to its charge, evaluates the state of the science for elucidating mechanisms of developmental toxicity and presents insights of developmental biology. It will show the promise of the subject in the next decade for understanding the action of developmental toxicants.

Advances in Developmental Biology

In the past 15 years, developmental biology has advanced remarkably, perhaps as at no other time in the field's history. It is now known that the trillions of cells of a large animal, such as a mammal, have the same genotype, which is the same as that of the single-celled zygote (the fertilized egg) from which the animal develops. That is to say, the genetic content of somatic cells does not change during the development of most animals. The recent clonings of Dolly the lamb, the Cumulina mouse family, and a nonhuman primate reaffirm the fact that a specialized cell, such as a mammary or cumulus cell, carries the genes for all other kinds of cells of the animal. The scientific advances that led to these clonings were built on earlier nuclear transplantation successes in frogs, first by Briggs and King, but particularly by Gurdon, which had led to similar conclusions for a nonmammalian vertebrate. Despite the same genes, the cells within the individual organism differ greatly in their appearance and functions, meaning that they have the same genotype and different phenotypes. The cell types differ greatly in the ribonucleic acids (RNAs) and proteins contained within them. They differ in which subset of genes they express from their total genomic repertoire. At least 300 cell types are recognized in humans (e.g., red blood cells, Purkinje nerve cells, and smooth or striated muscle cells). The number of cell subtypes is much larger, perhaps numbering tens of thousands, when further differences are taken into account related to the cell's stage of development and location in the body, as has been discovered in recent years. Development can be viewed as evolution's crowning example of complex gene regulation. From the single genome, thousands of different gene combinations must be expressed at specific times and places in the developing organism, and from the developing egg the information for the selective use of combinations must be generated.

A major factor in this regulation is the transfer of chemical information (i.e., signals) between cells during development. From recent research, which has built on earlier findings, the following is now realized:

- Embryonic cells of arthropods and nematodes make many of their developmental decisions based on which chemical signals they receive from other cells just as vertebrate embryonic cells do. Later the embryonic cells of all these organisms will make further decisions based on other signals. The cycles of signaling and responding are repeated over and over as development progresses. With that in mind and the fact that one genotype supports hundreds or thousands of cellular phenotypes, development can be said to rely on "genotype-environment interactions," where the local environment of each cell is generated by neighboring groups of cells. The genotype and cell's previous developmental decisions determine its options for responses to the signals currently present.
- The signaling pathways involved in this information transfer are known to be of 17 types (a few more may remain undiscovered). They are used repeatedly at different times and places in the embryo, from the earliest stages through organogenesis and cytodifferentiation, and even in the various proliferating and renewing tissues of the juvenile and adult.
- The signaling pathways are highly conserved across a wide range of phyla of animals (from chordates to arthropods to roundworms), presumably because they were present and already functional in the pre-Cambrian common ancestor of those animals.
- Many of the kinds of cell responses to signals also are conserved (e.g., responses of selective gene expression, secretion, cell proliferation, or cell migration). The response of developing cells to signals involves activation or repression of the expression of specific genes by transcription factors contained within genetic regulatory circuits. Signaling pathways frequently affect the activity of those factors. Many of the transcription factors and circuits are conserved across a wide range of phyla of animals.

Thus, an effective and general approach to the experimental analysis of developmental processes at all stages has been to inquire about the signaling pathways and transcriptional regulatory circuits that operate in the particular instance of development under study. Different organisms, which differ in aspects of their development, nonetheless use the same conserved signaling pathways and regulatory circuits, but in different combinations, times, and places, and have different genes as the targets of their transcriptional regulatory circuits. Processes of development, which seemed to confront scientists with infinite complexity and variety just a few years ago, now seem interpretable as composites of a small number of conserved elemental processes, namely, those of intercellular signaling, intracellular regulatory circuits, and a limited variety of targeted responses. These conclusions, which were reached by the analysis of development in animals as remote as mice, flies, and nematodes, give great validity to the use of model organisms in studying mammalian development, including that of humans, and in the future analysis of the action of developmental toxicants and in their detection.

Although the signal-response pathways are highly conserved, evolution has produced an increasing complexity of the "community" of pathways in vertebrates. This complexity is evident both in the increased number of closely related pathway components (diversifed protein family members) and in the increased possibilities for cross-talk among pathways. The redundant function of closely related components was made evident by existence of numerous targeted gene-knockout mutations in the mouse that produced little or no identifiable pheno-types—that is, the mice are normal or nearly normal under laboratory conditions. It must be emphasized, however, that functional redundancy provides two advantages. It protects the organism by ensuring that a fundamental process can proceed even in the absence or reduced presence of a critical gene activity. On an evolutionary scale, the multiplicity of overlapping functions provides a basis for generating diversity without losing essential functionality.

The Drosophila Breakthrough

The recent molecular understanding of developmental processes and components was gained from the experimental analysis of a few model organisms such as Drosophila melanogaster (the fruit fly), Caenorhabditis elegans (a free-living nematode), Danio rerio (the zebrafish), Xenopus laevis (a frog), the chick, and the mouse. D. melanogaster and C. elegans were chosen by researchers for their amenability to genetic analysis, afforded by their small size (hence, large populations) and short life cycle (hence, many generations). Nüsslein-Volhard and Wieschaus began a systematic search for developmental mutants of Drosophila in the mid-1970s. They submitted adults to high-frequency chemical mutagenesis and then inspected large populations of offspring for mutant individuals with strong and early developmental defects (before hatching) at discrete locations and discrete stages in the embryo. They discarded mutants with weak or pleiotropic effects as ones too difficult to analyze at their start. They examined mutagenized flies until the same kinds of mutants began to appear repeatedly in their collections. The recurrence was evidence that they had obtained all the different kinds of zygotic mutants (those affected in genes transcribed after fertilization) that mutagenized flies could yield under the conditions of inspection. This procedure is called "saturation mutagenesis," in which all the susceptible genes whose encoded products are important in development are thought to be revealed. Several laboratories, including those of Nüsslein-Volhard and Wieschaus, were also collecting maternal-effect mutants (those affected in genes transcribed in female germ cells before fertilization) and pursued this search to saturation.

The Drosophila mutants were categorized by phenotype and complementation behavior (putting two mutations together in a heterozygote to see whether they are alike or different) to establish the number of different genes whose mutations give the same phenotypic defect of development. Their categories included those embryos failing to develop the anterior or posterior end, odd or even segments, dorsal or ventral parts, mesoderm, endoderm, or nervous system. Further mutant combinations were made to establish epistasis (the interaction of different gene products, reflected in the dominance of one mutant defect over another) and to deduce plausible developmental pathways in which the actions of the encoded gene products could be related and ordered. By the late 1980s, a solid base of observations of Drosophila mutant phenotypes and gene locations had been built, and ordered pathways of function based on the mutant interactions had been proposed. This information served as the foundation for future molecular genetic analysis. The research was the first systematic and exhaustive approach to understanding an organism's development and to identifying components of developmental processes.

Synergy with Research Advances in other Areas

Meanwhile, other researchers worldwide made advances in biochemistry, molecular biology, cell biology, and genetics. They learned an enormous amount about the function of proteins in replication, transcription, translation, secretion, uptake, membrane trafficking, cell motility, cell division, the cell cycle, cell adhesion, and apoptosis (programmed cell death), to mention but a few of the cellular processes. Researchers improved the methods to isolate genes, sequence them, manipulate sequences, make transcripts in vitro, detect messenger (m)RNAs in cells by in situ hybridization, translate RNAs to proteins in vitro, and make antibodies to proteins. In situ hybridization, which graphically revealed the time and place of expression of specific genes in the embryo, was to prove particularly important for connecting the new molecular analysis to the older developmental anatomy. Much of the work was initially done with single-celled organisms: bacteria, yeast, or animal cells in culture. Some insights and techniques came from the study of cancer cells in the search for oncogenes.

In the course of that work, many of the processes, protein functions, and protein sequences were found to be strongly conserved among organisms as diverse as yeast and humans or even bacteria and humans. Various proteins of different organisms, and also within the same organism, shared "sequence motifs" by which the protein could be recognized as a member of a protein family with a particular function and descended from a common sequence ancestor. Newly discovered proteins could be assigned a function from just their possession of a particular motif. As more motifs were found, it will be easier to categorize newly discovered proteins. For example, receptor tyrosine kinases were recognizable by their transmembrane hydrophobic motifs and adenosine triphosphate (ATP)-binding domains. G-protein-linked receptors could be distinguished by a seven-pass (serpentine) transmembrane motif. Transcription factors could be recognized by the sequence motifs of their deoxyribnucleic acid (DNA) binding domains (e.g., zinc finger, basic helix-loop-helix, homeodomain, or leucine zipper domains). Of the recently sequenced genomes of yeast and C. *elegans*, for example, about 40% of the open reading frames (ORFs) are recognizable by known motifs. Function can be assigned, at least preliminarily, to the products of those genes. Plans are afoot to define the function of the missing ORFs of yeast and make the functions of all proteins assignable from sequence. At the same time, there are plans to identify a large number of protein-binding sequences in the regulatory regions of genes to be able to predict better the conditions of expression of genes.

Drosophila Development at the Molecular Genetic Level

By the time the Drosophila mutants were characterized in the mid-1980s, techniques were well-suited for molecular genetic analysis of affected genes and gene products. This part of the work moved quickly, thanks to gene-cloning techniques, background information about gene sequence motifs and protein function, and databases available to researchers worldwide. The successful isolation of a gene responsible for a developmental phenotype (when the gene was mutated) could be validated by the rescue of the mutant phenotype by transformation with the wild-type gene (usually as DNA included in a P-element transposon). In situ hybridization, coupled with color stains, readily revealed the normal time and place of expression of the specific genes whose mutations had been isolated. Regarding the function of these developmental genes, many were found to encode proteins with familiar motifs, such as those for receptor tyrosine kinases or various transcription factors. In fact, a surprisingly large number turned out to be transcriptional regulators. Function could be rapidly concluded from sequence data. Other Drosophila genes encoded proteins whose specific functions were unknown, yet they were recognizable generally as secreted proteins by their signal sequences or as new transcription factors by the fact they accumulated in nuclei and could bind to

DNA. In the course of this analysis, new intercellular signaling pathways were discovered, such as those involving the Decapentaplegic (DPP), Hedgehog (HH), Wingless (WG), and Notch/Delta ligands. (The whimsical names are those given by researchers to mutants based on the phenotypes.)

Hundreds of laboratories worldwide joined the work on Drosophila mutants, and the picture of early development took on a satisfying coherence and clarity, especially the steps of generation of segmentation and of the overall body organization in the anteroposterior and dorsoventral dimensions. These steps of early development are known collectively as "axis specification." The following is a brief summary of that picture to illustrate its completeness at the molecular level. The steps are stage-specific mechanisms of development. The mechanisms are now better understood in Drosophila than in any other organism. It is the kind of information scientists would like to have, but do not yet have, for mammalian development.

At the start of Drosophila development, the oocyte is provisioned with hundreds of maternal gene products that are uniformly distributed in the egg during oogenesis. Four gene products are spatially localized in the egg, however, and they provide the initial asymmetries on which the entire anteroposterior and dorsoventral organization of the embryo is built stepwise in development after fertilization. The four gene products include the following:

1. An mRNA located internally at the anterior end (encoding a transcription factor, named Bicoid).
2. An mRNA located internally at the posterior end (encoding an inhibitor of the translation of the mRNA for a transcription factor, named Nanos).
3. An external protein anchored to the egg shell at both ends of the egg (involved in the production of a ligand of a receptor tyrosine kinase in the egg-cell plasma membrane).
4. An external protein also anchored to the egg shell but at the prospective ventral side (involved in the production of a signal ligand of the Toll receptor in the egg-cell plasma membrane).

To exemplify the steps of use of those gene products, only one of the dimensions, the anteroposterior, will be described. The two mRNAs are initially at opposite ends of the egg. They are translated after fertilization, and the encoded proteins diffuse from the ends to form opposing gradients reaching to the middle of the egg. These proteins will act in concert to generate a gradient, high at the anterior end and low at the posterior end, of another transcription factor. The nuclear number increases rapidly in the uncleaved cytoplasm. The graded transcription factors, called members of the "coordinate class" or "egg-polarity class" of gene products, activate at least eight gap genes in nuclei along the egg's length at different positions, each position unique in terms of the local quantity of transcription factors of the coordinate class. (The terms "coordinate," "egg polarity," and "gap" also derive from mutant phenotypes.) The encoded gap proteins, which are all transcription factors themselves, accumulate in a pattern of eight broad and partially overlapping stripes along the egg's length. The proliferating nuclei are not yet separated by cell membranes—that comes later. These proteins in turn activate at least eight pair-rule genes, all of which also encode transcription factors. Complex cis-regulatory regions of the various pair-rule genes define their expression responses to the spatially distributed gap proteins. The pair-rule proteins then activate at least 12 segment-polarity genes, some of which encode transcription factors and some of which encode secreted protein signals. The pair-rule and gap proteins together also activate eight homeobox (Hox) genes to be expressed in broad stripes. Thus, the early steps of development involve cascades of transcription factors distributed in space according to the initial gradients of a few agents and to the expression rules contained in the complex cis-regulatory regions of genes for yet other transcription factors. These key steps are accomplished in the first 3 hours of development, mostly before cell membranes are formed and gastrulation begins, although the final elaboration of the segment-polarity and Hox genes occurs after cells form.

Once the segment-polarity genes and Hox genes are activated, they maintain their expression in cells by an auto-activating circuitry, in some cases by the encoded transcription factor activating expression of its own gene. The coordinate, gap, and pair-rule proteins are then no longer needed. Their products disappear, and the genes are no longer expressed.

Similar conclusions apply to the development of the termini and the dorsoventral dimension, which also rely on initially asymmetric signals. The developmental mechanisms of the termini and dorsoventral dimension are of additional interest, because the signals bind to transmembrane receptors and activate signal transduction pathways, eventually leading to the activation of transcription factors and new gene expression. These inductions are the first to occur in the developing Drosophila egg. Approximately 100 genes and encoded gene products have been identified as necessary to establish the organization of the early gastrula. Hundreds more participate in the accomplishment of these events, but they are less well described at present. In most cases, these genes probably encode proteins required in numerous developmental processes and, hence, were not recovered under the conditions of the mutant inspections used here.

As shown in figure below, a coherent scheme of early development was proposed and well supported by 1992, the first of such complexity and completeness at the molecular level for any organism.

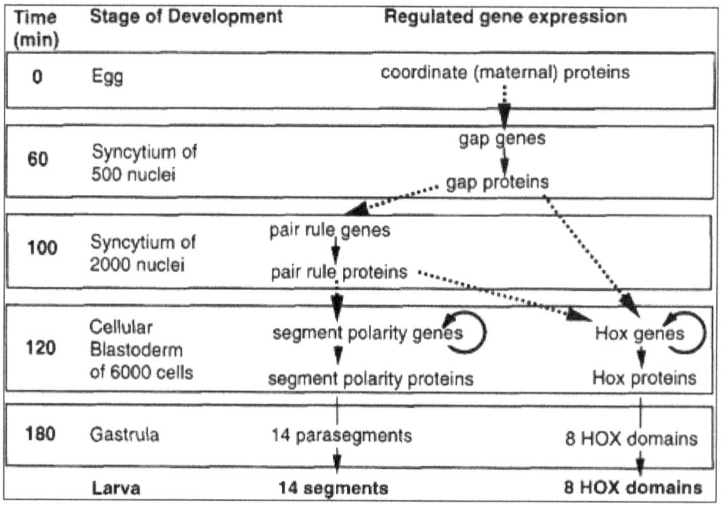

In figure, Outline of anteroposterior development in Drosophila and the steps of regulated gene expression. Heavy dashed arrows indicate the activation of specific gene expression by transcription factors. Thin solid arrows indicate transcription and translation. Note that Hox genes are activated by both pair-rule and gap proteins, whereas segment-polarity genes are activated by pair-rule proteins alone. In the anteroposterior dimension, segments and HOX domains are formed.

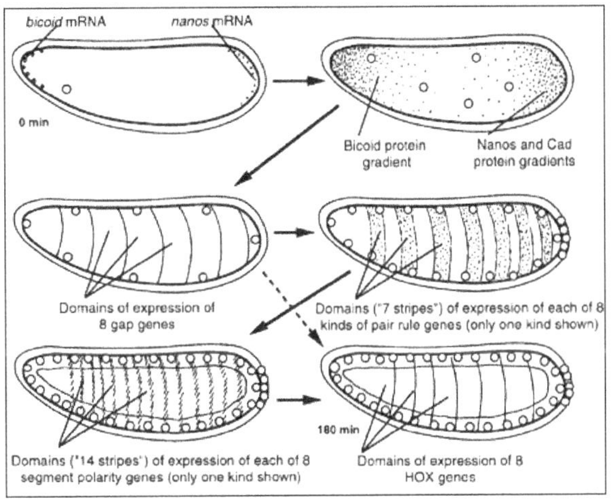

Anteroposterior development in Drosophila.

In figure is shown diagrammatically here, for segment formation and HOX compartment formation. The coordinate proteins Bicoid, Nanos, and Cad are translated from mRNAs localized at the two poles of the egg during oogenesis. Translation generates gradients of proteins. Bicoid and Cad are transcription factors, whereas Nanos protein inhibits the translation of another translation factor (Hunchback) in the posterior half of the egg. The graded transcription factors activate eight gap genes, and different factor concentrations activate different gap genes. The gap proteins are also transcription factors. Each diffuses locally and inhibits other gap genes, setting up eight partially overlapping stripes of gap protein along the egg's length. The gap proteins activate eight pair-rule genes, each of which has a complex cis-regulatory region and is activated by seven combinations of gap proteins, each making seven evenly spaced stripes of protein. Thus, there are 8 × 7 or 56 stripes of pair rules along the egg's length, arranged in 7-fold repeats. The pair-rule proteins are all transcription factors. These activate eight segment-polarity genes, each of which has a complex cis-regulatory region activated by at least two combinations of pair-rule proteins, to give 14 stripes of expression each. Thus, there are 14 × 8 or 104 stripes of segment-polarity proteins. The 14-fold repeat is the basis for 14 segments of the posterior head, thorax, and abdomen. The pair-rule and gap proteins together activate Hox genes in eight domains in the posterior head, thorax, and abdomen. Cell outlines are not shown, but cells are present in the two lowest panels.

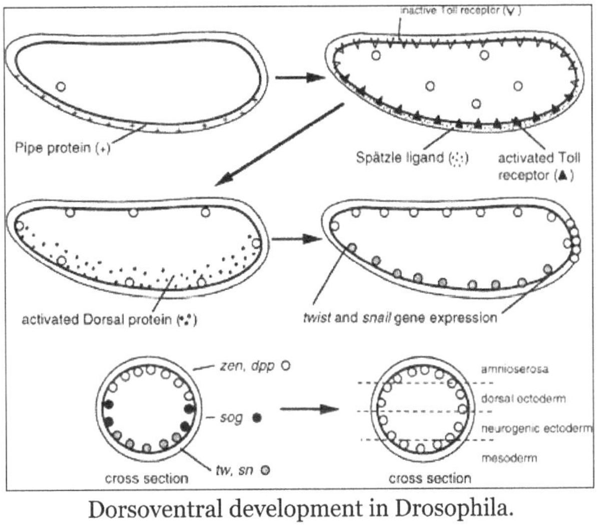

Dorsoventral development in Drosophila.

In figure, the egg shell contains Pipe protein on the future ventral side, deposited there during oogenesis. After fertilization, the egg secretes several proteins into the space between the egg shell and plasma membrane. Pipe activates one of the proteins, which then sets off others in a protease cascade, the last member of which cleaves the Spätzle protein, releasing a ligand that binds to the Toll transmembrane receptor, which is uniformly distributed over the egg surface but ligand-activated only on one side. The activated receptor, via several intracellular steps, activates the Dorsal protein, a transcription factor, which enters local nuclei and activates two genes, Twist and Snail, which also encode transcription factors. Those activate other genes for gastrulation and for mesoderm formation on the ventral side. Thus, the Pipe protein is involved in a kind of mesoderm induction. Active Dorsal protein also represses the Zen and Dpp genes on the ventral side. On the dorsal side, Dorsal protein remains inactive and the Zen and Dpp genes are expressed. Laterally, there is enough active Dorsal protein to repress Zen and Dpp but not enough to activate Twist and Snail. Here, the Sog gene is permissively expressed and not repressed, preparatory to neurogenic ectoderm formation. Thus, the dorsoventral dimension of the egg is divided into three domains of gene expression. Later, the Sog protein is secreted and diffuses to the Zen, Dpp region, inhibiting Dpp signaling and allowing the division of that region into two subregions the prospective amnioserosa and prospective dorsal ectoderm.

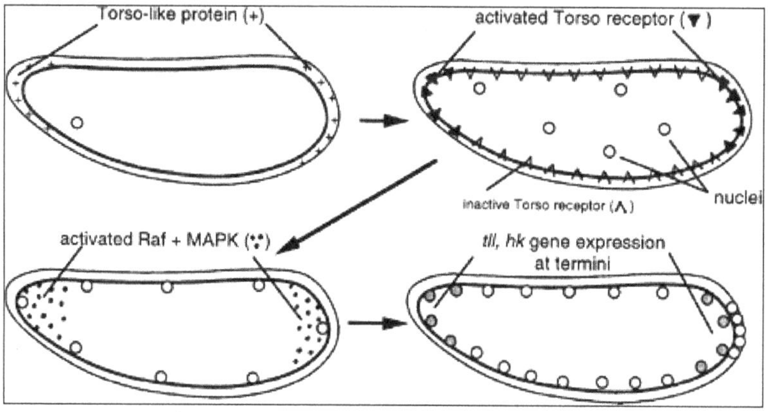

The development of termini in Drosophila

In figure, The Torso-like protein is present in the egg shell at the two ends of the egg, deposited there during oogenesis. After fertilization, the egg secretes several proteins into the space between the plasma membrane and egg shell. The proteins include proteases that are locally activated at the end by the Torso-like protein and release a ligand that binds locally to the transmembrane Torso receptor, a member of the RTK signal transduction family. The activated receptor locally activates Raf and MAPK, which phosphorylate a transcription factor locally, inhibiting its repression of genes and allowing local expression of the Tailless (Tll) and Huckebein (Hkb) genes involved in formation of the endoderm, terminal ectoderm, and gut involution during gastrulation. Thus, the Torso-like protein is involved in endoderm induction.

Hox Genes and the Drosophila Connection to Vertebrate Development

Even though researchers in other areas widely appreciated the breakthroughs in Drosophila development, they questioned the relevance of the information to vertebrate development. Vertebrates, as chordates, were thought to have branched from arthropods long ago and last shared a very simple common ancestor in the pre-Cambrian era (about 540 million years ago). The two groups were thought to have evolved their segmentation and heads independently. One of the first significant similarities between vertebrate and fly development came from work on homeotic genes, now called Hox genes. As mentioned before, the Hox genes are expressed in eight broad bands or spatial compartments in the anteroposterior dimension of the body shortly after gastrulation but prior to organogenesis and cytodifferentiation. Their encoded products make each spatial compartment different from the others.

The study of the eight Hox genes of Drosophila was primarily pioneered by E. Lewis from 1940 to 1970. For his work in that area, he shared the Nobel Prize with Nüsslein-Volhard and Wieschaus in 1995. Lewis selected Drosophila mutants that exhibited mislocated body parts (e.g., wings in place of halteres (balancing organs) and legs in place of antennae). The term "homeotic" connotes such mislocation without distortion. In the homeotic mutant, the anteroposterior dimension of the animal has fewer anatomical differences along its length. For example, the Ubx mutant has an extra mesothorax located at the normal metathorax position but lacks a metathorax. It has four wings but no halteres, whereas normal Drosophila have two wings and two halteres. When the first two Hox genes (Ubx and Antp) were isolated, their sequences were compared and a shared 60-base sequence was found, the homeobox. The sequence is the same in both genes except for a few bases. That sequence encodes the DNA-binding motif of the encoded proteins, which are members of a large and ancient family of transcription factors. The other six Hox genes were soon isolated from Drosophila, and those too had closely related homeobox sequences. Then the eight genes were shown to exist in a contiguous cluster (actually two subclusters in D. melanogaster but one in another arthropod, Tribolium), probably all tandemly duplicated and diverged from a few founder sequences in an ancestor of arthropods. Furthermore, the members are expressed in stripes in the anteroposterior dimension of the body, in an order identical to their gene order on the chromosome (a correspondence referred to as "colinearity" of gene order and expression).

In the mid-1980s frogs and mice were found to contain similar sequences, also arranged in contiguous gene clusters. Interestingly, their expression in mice showed the same anteroposterior colinearity as that in Drosophila. As an evolutionary explanation, the common ancestor of arthropods and chordates must have had a complex Hox cluster already functioning in its development. Vertebrates, however, differ from arthropods in having at least four multi-member clusters instead of one. A comparison of gene arrangements and domains of expression in Drosophila and mammalian (mouse) Hox clusters.

Such genes are called selector genes because their encoded products, which are transcription factors, select which other genes will be expressed in that spatial compartment of the body. The thousands of target genes of a selector-gene product encode proteins involved in subsequent local development, including the many kinds of organogenesis of different parts of the body. Hox genes have a central role in development. Because of them, the coordinate, gap, and pair-rule proteins of early development do not have to directly activate those thousands of target genes in a region-specific way but activate only the Hox genes, whose encoded proteins then do the job of regulating sets of genes in their respective regions. Methods for the directed knockout of genes in mice were invented by the mid-1980s as a way to test gene function, and the Hox genes of mice were found to control aspects of local development in their compartments, especially in vertebrae, neural tube, and neural crest derivatives. Their selector role was similar to that in Drosophila. However, many of the target genes of Hox proteins in mice and flies are clearly different.

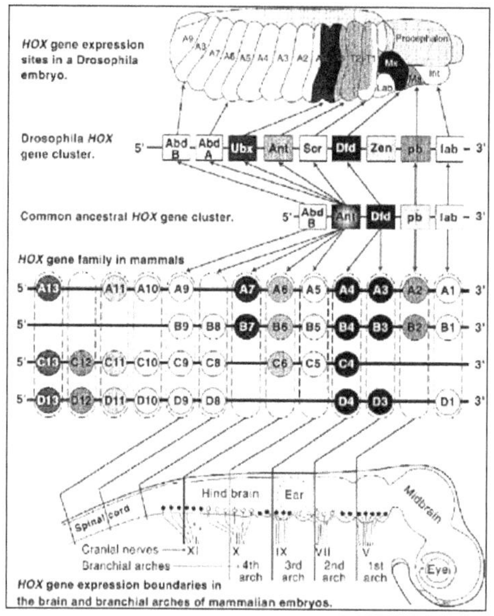

This figure illustrates the striking similarities of gene organization and expression of Hox clusters in Drosophila and mammalian (mouse) embryos. At the top is a 10-hour Drosophila embryo showing expression zones of individual Hox genes in thoracic (T1-3) and abdominal (A1-9) segments and parts of the head (Lab, labrum; Mx, maxillary; Ma, mandible; Int, intercalary segment). Note the colinearity of Hox gene expression sites along the anterior-posterior body axis to their 3′ to 5′ location along the chromosome. The greatly expanded vertebrate Hox gene family is shown in the middle. These genes are arranged in four clusters (labeled A, B, C, D), each on a separate chromosome. Having arisen by duplications early in chordate evolution, Hox genes in paralogous groups (e.g., A4, B4, C4, D4; shown enclosed in dashed boxes) are more closely related than are adjacent genes (e.g., B3 vs. B4 vs. B5). The four most 5′ paralogous groups have no close equivalent in arthropods; these are expressed in the tail and fins or limbs. Lines extending from each paralogous group to the schematic brain and cranial spinal cord show the rostral limits of expression of members on each group. Note, again, the colinearity between expression sites and relative chromosomal position of most Hox genes. The same is generally true for somites and, in the proximo-distal orientation, for limbs.

The Hox clusters of Drosophila and chordates are under intense study. It is now known that genes of four mouse clusters are coordinated in an elaborate circuitry of auto- and cross-activation and repression, in which the genes near the 5′ end of the DNA sequence tend to repress genes near the 3′ end when both are initially expressed in same cell. Equivalent paralogs in different clusters tend to overlap in the target genes they activate and repress, but each has some unique targets, as shown by the phenotypes of single-Hox knockout mutants of the mouse. As a whole, the Hox genes operate as a complex genetic regulatory system rather than as independent members.

More recently, the Hox-like Ems and Otd genes have been discovered in Drosophila as expressed in the head in regions anterior to the expression compartments of the Hox genes. Homologs of these genes (called Emx and Otx) have been found expressed in the head of the frog and mouse anterior to the Hox gene domains of the posterior head, thorax, and trunk. This was a surprise, because evolutionary biologists had thought that the vertebrate head is unique to that group and has little in common with the head of a common ancestor of vertebrates and arthropods. However, even that complexity of body organization, like HOX compartments, must predate the branching of arthropods and chordates.

The Emergence of Caenorhabditis Elegans

The free-living nematode Caenorhabditis elegans emerged as an important model system in the 1970s, as the result of pioneering work on its genetics by S. Brenner. Chosen for its short life cycle (3 days) and general amenability for genetic analysis, small size (1-mm length), transparency, and simplicity (only 959 somatic cells), C. elegans quickly attracted a following among developmental biologists and geneticists. In particular, J. Sulston was primarily responsible for first describing the complete cell lineage from fertilization to adulthood and then spearheading the physical mapping and DNA sequencing of the genome. C. elegans recently became the first metazoan organism whose genome is completely sequenced. In the meantime, researchers from many laboratories isolated mutants and identified many important genes controlling development, the result being that C. elegans is now the most completely described and one of the best understood models for development. In some ways, the development of vertebrates is more similar to that of C. elegans than of Drosophila (e.g., having a cellular rather than a syncyctial early embryo), and in other ways less similar (e.g., having a highly invariant cell lineage and a fixed small number of cells, no Sonic Hedgehog signaling pathway, and few HOX genes). These two model animals complement each other usefully for research into fundamental mechanisms of metazoan development.

Conserved Developmental Processes

Researchers increasingly suspected similarities of development between fruit flies and mice and began to look systematically for homologs of Drosophila developmental genes in mice, frogs, and chicks. In the late 1980s, this was a new research approach. Its success has favored the impression that at a gross level, nematodes, flies, and mice are “all the same organism” and that what is learned about one will have relevance to the others. In a genetically tractable organism, such as Drosophila or C. elegans, a gene is isolated by using a screen for a particular kind of developmental failure, and then the role of its encoded product in development is efficiently deciphered in that organism. Homologs of “developmentally interesting” genes are then sought in vertebrates, such as mice or frogs, in which mutant searches are still daunting due to the comparatively small populations and slow development. The homolog's function is thereafter studied in the vertebrate, for which the Drosophila or C. elegans information is used as a guide. The mouse is attractive for such studies, because the homologous gene can be knocked out and the phenotype of the null mutant examined to learn about the function of the encoded product.

A surprising array of developmental components and processes is shared between Drosophila and vertebrates (i.e., between arthropods and chordates). In addition to the EMX, OTX, and HOX organization of the body plan, they share the compartments of the dorsoventral dimension (which are thought to be inverted in orientation in one group relative to the other); the presence and mode of organogenesis of limbs (appendages), eyes, heart, visceral mesoderm, and gut; the steps of cytodifferentiation during neurogenesis and myogenesis; and even segmentation. Although the anatomical structures themselves are very different between arthropods and chordates, a number of the underlying steps of development are the same. The last common ancestor of chordates and arthropods was, it seems, a pre-Cambrian animal of much greater complexity than previously realized. Divergent groups of metazoa (members of the animal kingdom) can be treated as “the same organism” in the experimental analysis of many fundamentals of development. From all of those similarities, the value of model systems for gaining an understanding of difficult basic problems in mammalian development, including that of humans, is undeniable. Humans, flies, and even roundworms are less different than widely thought just 10 years ago.

Signaling Pathways in Development

An important realization to come from the Drosophila research concerns the pervasive use of cell-cell signaling in most aspects of development, starting with the termini and dorsoventral dimension and extending to organogenesis of many kinds. Inductive signaling was thought to be important in vertebrate development, as mentioned above, but insects and other invertebrates had been assumed to develop as composites of independent lineages of cells ("mosaic" development). This is not at all the case. Six signaling pathways are used repeatedly in early Drosophila development: the Hedgehog, Wingless-Int (Wnt), transforming growth factor β (TGFβ), Notch, receptor tyrosine kinase (RTK), and cytokine receptor (cytoplasmic tyrosine kinase) pathways. Comparative studies soon showed that these pathways exist in vertebrates as well, and most also exist in nematodes (except the Hedgehog pathway). Four other conserved pathways in addition to those six are used heavily in later development, mainly in organogenesis, and seven others come into use in the physiological functioning of the organism's differentiated cell types. The number of known pathways has now reached 17. Each pathway is distinguished by its unique set of transduction protein intermediates.

Table: Similarities of Arthropods and Chordates.

Developmental Process	Conserved Genes	Organisms That Share Process
Anteroposterior organization	Hox gene complex: similar order of genes in the cluster and similar order of expression domains in the posterior head and trunk (thorax and abdomen).	Drosophila and mouse
Anterior head organization	Ems-Otd (Emx-Otx) selector genes: similar nesting expression domains in the anterior head.	Drosophila and mouse
Dorsoventral organization	Sog-Dpp-Tolloid (Chordin-BMP2,4-xolloid): similar gene expression domains, similar protein interactions in the neural versus epidermal regions; similar gene expression domains in the visceral mesoderm and heart. Was the chordate dorsoventral axis formed by inverting the axis of an arthropod ancestor?	Drosophila and Xenopus
Segmentation	*Engrailed* and *HH-SHH* expression domains are similar in posterior half of segment or somite; *Hairy* gene expression in alternate segments or somites.	Drosophila, amphioxus, and zebrafish
Appendage or limb patterning	Similar domains of WG-HH-DPP (WNT-SHH-BMP) signaling and expression of En, Ap (En, Lmx) selector genes	Drosophila, chick, and mouse
Eye specification	Similar domains of expression of *Eyeless-Pax6* and *Sine oculis-eye*selector genes.	Drosophila, mouse, and human

Although the organisms of these two phyla seem very different (e.g.,insects and crustaceans versus fish and mammals), they share many developmental processes at the level of their use of combinations of signaling pathways and genetic regulatory circuits. In italics are various similar conserved genes used in the conserved processes. These similarities serve as evidence that the pre-Cambrian common ancestor of chordates and arthropods was already complex in its anteroposterior and dorsoventral organization and perhaps segmented. Many aspects of cytodifferentiation are also similar (e.g., the use of MyoD in muscle and Achaete-Scute in nerve cells).

As a generalization, most of the pathways involve transmembrane receptor proteins that bind ligands at the extracellular face, as diagramed in figure. Ligands arrive in some cases by free diffusion after secretion from distant neighbor cells. Others diffuse only short distances or remain attached to the surface of the cell of origin, reaching only the contacting cells. Activated receptors of the recipient cell activate the first intracellular component of a signal transduction pathway, and this then activates a subsequent component, and so on. Some pathways are long, with 7-10 intermediates. Others have one or two. The nuclear hormone receptor pathway is the shortest, having only one step. In this case, hydrophobic ligands penetrate the cell membrane on their own and bind to a receptor protein, which also functions as a transcription factor. In the longer pathways, a change of activity is passed along a series of on-off switches, which constitutes an information relay pathway, or signal transduction pathway. Ultimately, in some pathways, a protein kinase is activated at the end of the series, and that enzyme phosphorylates numerous target proteins, which change their activity (activated or inhibited) because of the phosphate addition. The target proteins are components of various basic cell processes, such as transcription, the cell cycle, motility, or secretion. Hence, these processes are turned on or off, and the change of function constitutes the cell's response to a signal. In many other pathways, a specific transcription factor is activated at the end of the pathway, and this factor is a pathway component. In development, the most frequent target of signaling pathways is indeed transcription. The pathways used in early development tend to have transcription as the only target. That is, particular transcription factors are phosphorylated or proteolyzed as a signal transduction step of the pathway, changing their activity in activating or repressing particular genes.

The pathways are used repeatedly at different times and places of development in Drosophila, nematode, and vertebrates, as listed in table. Drosophila null mutants are usually lethal if they lack a step in any of those pathways. Lethality is an indication of the essentiality of those signaling functions. However, in the mouse (and probably all vertebrates), a null mutant for a step of a pathway is often not lethal. The mutant is born with a limited abnormality of anatomy and sometimes of behavior. The genetic basis for some of these mouse developmental defects is fully known. In some cases, the mutant mice live to adulthood and reproduce.

Table: The 17 Intercellular Signaling Pathways.

Period During Development	Signaling Pathway Used
Early development (before organogenesis and cytodifferentiation) and later (during growth and tissue renewal).	1. Wingless-Int pathway. 2. Transforming Growth Factor β (receptor serine and threonine kinase) pathway. 3. Hedgehog pathway. 4. Receptor tyrosine kinase (small G proteins) pathway. 5. Notch-Delta pathway. 6. Cytokine receptor (cytoplasmic tyrosine kinases) pathway (STAT pathway).

Middle and late development (during organogenesis and cytodifferentiation) and later (during growth and tissue renewal).	7. Interleukin-1-Toll Nuclear Factor-Kappa B pathway. 8. Nuclear hormone receptor pathway. 9. Apoptosis pathway. 10. Receptor phosphotyrosine phosphatase pathway.
Larval and adult physiology (after cell types have differentiated).	11. Receptor guanylate cyclase pathway. 12. Nitric oxide receptor pathway. 13. G-protein coupled receptor (large G proteins) pathway. 14. Integrin pathway. 15. Cadherin pathway. 16. Gap junction pathway. 17. Ligand-gated cation channel pathway.

The pathways are shared by most animals (metazoa). Note the six pathways used heavily in the early development of most animals (i.e., at stages before organogenesis and cytodifferentiation begin). Along with those pathways, four more are used in later development in the periods of organogenesis and cytodifferentiation, growth, and tissue renewal. Seven pathways are used mostly in the physiological function of differentiated cells; much of that function also involves signaling. Each pathway is identified by the particular transduction intermediates it contains. Most pathways are unique to metazoa, although components of each are often found in single-celled eukaryotes in other signaling roles, such as pathways of checkpoint control, stress response, infection response, mating, and feeding.

Are the signaling pathways less important in vertebrate development than in Drosophila? No, the nonlethality in vertebrates reflects the fact that the pathways have a substantial redundancy of signaling components. It is postulated that early in vertebrate evolution (as jawless fish arose), the genome underwent a quadruplication. In addition, some genes underwent tandem duplications. Many of the initially identical genes later diversified their sequences, leading either to diversified functions of encoded products or to diversified cis-regulatory regions setting different conditions of expression. Despite this diversification, extensive redundancy still remains. For example, there are more than 24 TGFβ ligand genes and 11 Wnt ligand genes in mouse, but only 3-5 TGFβ and 1-3 Wnt genes in Drosophila. In mouse development, the various genes for a single step of a single kind of signaling pathway are expressed at different times and places in the embryo. When one gene is knocked out, the defect in development is limited in scope to a few times and places where no related gene is expressed to provide overlapping function.

In figure, a generalized signal transduction pathway (information transfer via a series of on-off switches. The active ligand is shown on the left, approaching the transmembrane receptor protein. Inside the cell is a multistep signal transduction pathway composed of switch-like intermediates, A, B, and C, that can exist in active and inactive states. In the absence of signal, the intermediates are inactive. During binding, the receptor becomes active and activates one intermediate that activates the next, and so on in series, until eventually a protein kinase is activated. This pathway transfers information, not energy or materials. The kinase enzyme is specific for transferring phosphate from adesine triphosphate to a serine, threonine, or tyrosine residue of select protein targets within the cell. Phosphorylated target proteins change activity, becoming either active or inactive, and therefore the basic cellular process of which they are a part, changes activity. In this way, the signal has effects (i.e., it triggers cellular responses). Target proteins might be components of the processes of transcription, translation, the cell cycle, cell movement, differentiation, or other signaling pathways. Transcription is a particularly frequent target. Fourteen of the 17 involve transmembrane receptors; the other two involve intracellular receptors and the ligands pass through the plasma membrane readily.

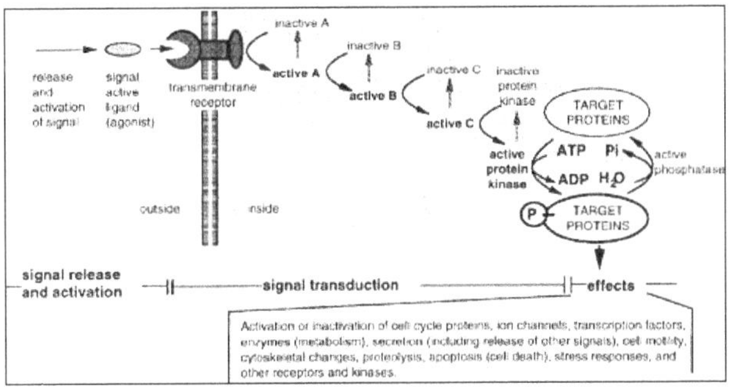

Table: Various Uses of Signaling Pathways in Vertebrate Development.

Pathway	Use
Wnt pathway	Dorsalization of the body (fish, frogs) induction of the organizer or node; posteriorization of the neural plate, midbrain development; dermamyotome induction, somite dorsoventral organization; dorsalization of the fin or limb; female reproductive development; kidney development; dorsoventral differences of limb, hematopoiesis.
Transforming growth factor β pathway (BMP,TGFβ, GDF, VEGR, Nodal, Activin, Dorsalin).	Mesoderm induction; induction of organizer, left-right asymmetry; ventralization of mesoderm and ectoderm; neural-crest development to neurons; chondrogenesis of limb, bone development; digit-web spacing, tooth, heart; notochord induction of floor plate of neural tube; notochord and floor-plate induction of sclerotome of somite; prechordal mesoderm induction of prosencephalon; left-right asymmetry, somitogenesis, lung; fin or limb development; zone-of-polarizing-activity (ZPA) induction of anteroposterior axis; gut and visceral mesoderm, hair follicle, skin, and tooth development; spermatogenesis
Hedgehog pathway (Sonic, Indian, Desert).	Notochord induction of floor plate of neural tube; notochord and floor-plate induction of sclerotome of somite and the dorsoventral organization of neural tube; prechordal mesoderm induction of prosencephalon; inhibit cyclopia; ZPA induction of anteroposterior axis of fin or limb development; gut and visceral mesoderm development; hair development.
Receptor tyrosine kinase pathway (EGF, FGF, PDGF, EPH).	Mesoderm maintenance; limb (apical ectodermal ridge), vasculogenesis; hair follicle, inner ear, retinotectal projection; astrocyte differentiation, branchial arch signal to neural crest; heart, lung, and tooth development.
Notch-Delta pathway (Delta, Serrate, Jagged).	Several steps of neurogenesis; oligodendrocyte differentiation, retina development; somitogenesis, inner-ear development; feather-bud development; blood-cell development (e.g., thymocytes).
These pathways are used repeatedly in early development. The particular uses in developmental processes are well conserved across vertebrates.	

Antagonists of Signaling

Although only a small number of signaling pathways operate in early development, these have several regulatory features of relevance to developmental toxicity. One is that many of the pathways engage in "cross-talk" with one another, so that activation of one enhances or suppresses the activation of another. Cross-talk occurs at all levels, from effects on ligand availability to effects on target function. For example, the active RTK pathway can lead to phosphorylation and activation of GSK3 in the Wnt pathway, or phosphorylation and inactivation of a SMAD protein of the TGFβ pathway. Signaling via one pathway, such as the SHH pathway, can lead to repression of gene expression of the components of another pathway, such as the Nodal TGFβ pathway. Signaling via the RTK or Wnt pathways can negatively affect signaling via the Notch pathway.

A second regulatory feature is the production by cells of antagonists of signaling. For example, embryonic cells can produce proteins, such as Chordin or FRZB, that bind directly to the BMP4 or WNT8 ligands, respectively, and block their capacity to bind to their receptors. Hence, signaling is prevented even though the signal is present. The Chordin protein is further regulated by a protease that degrades it, an antiantagonist. Thus, the pattern of activity of a signaling pathway is subject to extensive modulation, both positive and negative. Several kinds of adjacent cells can affect the outcome of signaling. It is plausible that some of these antagonist proteins are targets of toxicants. Disrupting the activity balance of agonists and antagonists in a region of the developing embryo would be expected to disrupt development, leading to the over- or under-development of an organ. That could occur without a change in level of the agonist or antagonist.

Finally, in many pathways, the operation of the pathway leads to the generation of self-inhibiting components in the cells receiving signals, a feedback mechanism, which is thought to have importance in the controlled spatial responses of cells to diffusible signals. Toxicants upsetting these feedbacks would upset development.

Molecular-stress Pathways and Checkpoint Pathways

Molecular-stress pathways and checkpoint pathways are not pathways of intercellular signaling but of intracellular signaling, reflecting an individual cell's sensing of disruptions of normal cell function and development due to either physical or chemical agents of the environment (the molecular-stress pathways) or the cell's own internal imbalance or errors in its synthetic activities (the checkpoint pathways). These pathways are widely present in single-celled eukaryotes (e.g., yeast) and even prokaryotic cells, as well as in animals. Several molecular-stress and checkpoint pathways are listed in table.

Table: Natural Antagonists of Signaling Pathways.

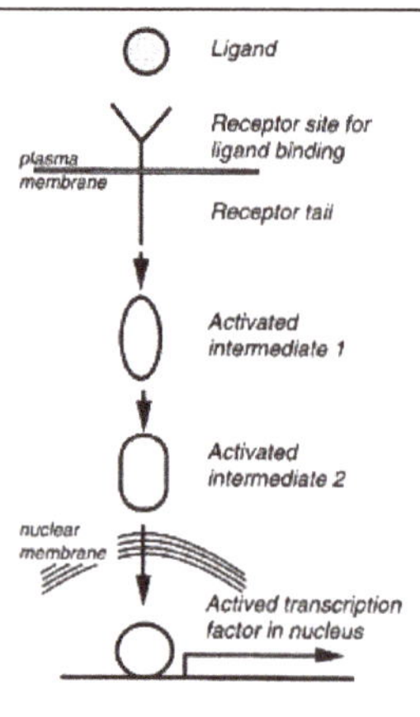

	TGF-β pathway Antagonists	Wnt Pathway Antagonists	RTK Pathway Antagonists	Hedgehog Pathway Antagonists	Notch/Delta Pathway Antagonists
	Noggin, Chordin, WIF1, Cerberus, Dickkopf, Xnr3	Frzb, Cerberus, Sizzled	Inactivity of Rhomboid		Fringe
	Inactivity of one-eyed pinhead		Kekkon 1, Argos	HIP, Patched	Wnts, EGF
	Smad 7, 8				Dishevelled, Inactivity of Presinillin
	MAPK?	Axin	Sprouty, Puckered, Erp phosphatase	PKA	
		Nemo-like kinase			
Brinker					Hairless

Cell-cell signaling in metazoa involves antagonists as well as agonists, making signaling more complex and conditional. Antagonists can act at any of several steps in a pathway. Those acting at the ligand or receptor site levels are secreted and affect many nearby cells. Those acting at intracellular steps affect only the cell in which they are produced.

Molecular-stress pathways are activated when the cell suffers some chemical alteration, such as damage to DNA (e.g., by X-ray, UV, or alkylating agents) or denaturation of proteins (e.g., by hyperosmotic conditions, oxidation, heat, or alcohol). The cell's signaled response is one of repair and homeostatic counteraction. In the case of the cytosolic unfolded protein pathway (previously called the "heat shock response"), chaperone proteins, such as Hsp90, help to refold denatured proteins, restoring their activity. These same chaperones play a folding role in the normal synthesis and deployment of intrinsically unstable proteins, such as cell-surface receptors. Recent experiments have suggested that if Hsp90 is partially disabled by mutation or overloaded by stress, variant proteins in some members of a population might be unable to fold correctly, resulting in developmental defects. Another example would be the multidrug transport proteins (P-glycoprotein of mammals) that are induced in the presence of high drug levels and serve to export a wide variety of drugs from the cell.

In checkpoint pathways, the cell's response is one of delaying certain synthetic processes until other processes are complete. These controls are important in coordinating the timing and extent of cellular processes, such as ensuring the completion of DNA synthesis before mitosis begins or ensuring the attachment of chromosomes to the spindle before anaphase begins. An example of relevance to developmental toxicology is that when colchicine (a Vinca alkaloid) inhibits microtubule formation in a mitotic cell, the cell is prevented from initiating anaphase because of a checkpoint control pathway, which assesses the attachment of kinetochores to spindle microtubules. While chromosomes remain unattached, anaphase is not initiated. When the inhibitor is removed, the cells assemble a spindle and proceed with anaphase. Mutant cells have been isolated that lack components of the control (such as the MAD_2 protein), and these cells initiate anaphase in the presence of colchicine, without a spindle. They suffer extensive aneuploidy.

Checkpoint and molecular-stress pathways work together. For example, damaged DNA inside the cell triggers various stress pathways, leading to DNA repair. During the repair, the checkpoint pathways delay DNA synthesis or mitosis until repair is complete. In the context of this committee's evaluation, there are two relevant points about those widely distributed pathways:

- They offer possibilties for the detection and analysis of developmental toxicants, because they indicate the cell's state of stress in the presence of toxicants. Broadly acting toxicants are likely to show up as triggers of stress responses.
- The intercellular signaling pathways of metazoa probably arose in evolution as elaborations and reworkings of the more ancient molecular-stress and checkpoint pathways of single-celled eukaryotic ancestors. This is surmised because a number of the intermediates (e.g., protein kinases) of the molecular-stress and checkpoint pathways are also used in the metazoan signaling pathways.

Table: Molecular-Stress Response and Checkpoint Pathways.

Checkpoint Pathways	Function and Cell Response
G1/S checkpoint	Monitors nutritional state, biosynthetic capacity, and cell adhesion and imposes G1 arrest until cell is prepared for S phase.
G2/M checkpoint	Monitors completion of DNA synthesis (S phase) and imposes G2 arrest until cell is prepared for M phase.
Metaphase/anaphase checkpoint	Monitors attachment of chromosomes to the spindle and imposes metaphase arrest until cell is ready for anaphase.
Molecular-Stress Response Pathways	Function and Cell Response.
DNA damage (genotoxic stress)	Kinases are activated at DNA damage site by single stranded DNA and 5′, 3′ ends, leading to p53 activation and transcription of genes encoding p21 inhibitors of cyclin-dependent kinases, and hence G1/S or G2/M arrest until repair is complete.
Cytosolic unfolded protein pathway	Activated by heat, alcohol, anaerobiosis, and amino acid analogs, leading to activated transcription of genes encoding chaperone proteins until protein refolding is complete.
Endoplasmic reticulum (ER) unfolded protein pathway	Unfolded proteins activate receptor thre/seri kinase, leading to release of a nuclease that degrades some mRNAs and reduces translation (G1 arrest), and splices some mRNAs leading to transcription of genes encoding chaperone proteins until protein refolding is complete.
Apoptosis (cell death)	Triggered by intracellular damage or extracellular signals, leading to caspase protease activation and cell destruction.
Ultraviolet, hyperosmotic shock, free-radical oxidation pathways	Mediated by MAP kinases, leading to transcription, until damage is reversed.

Many of these are found in single-celled eukaryotes as well as in most or all cells of animals.

One more kind of molecular-stress and checkpoint pathway should be noted: the apoptosis pathway, which is also a signaling pathway. It can be activated by either extracellular or intracellular signals and leads to the "programmed" death and destruction of a cell. It is a tightly controlled process in which a cell is destroyed but neighboring cells are unaffected. Apoptosis is not found in single-celled organisms. It is an invention of metazoa and is used in normal embryonic development as well as in recovery attempts of teratogen-damaged embryos. In normal development, where apoptosis is also known as programmed cell death, it is important in the shaping of tissues and organs (e.g., the elimination of cells from the interdigital spaces of the human hand). Cells undergoing apoptosis are found throughout most embryonic mesenchymal tissues, presumably reflecting the elimination of cells that have not been able to successfully integrate the signals impinging upon them. Some mouse mutants, such as Hammertoe, fail to initiate the normal amount of programmed cell death in normal limb development, and an abnormal limb results.

Apoptosis is also the ultimate molecular-stress and checkpoint pathway, for it eliminates cells too damaged to be restored to a normal state by the various repair and checkpoint pathways. For example, if DNA repair is incomplete and the cell attempts to divide, it is killed and autolyzed. It has been proposed that cell death is less detrimental to the multicellular organism than having live cells with highly modified DNA, perhaps proliferating uncontrollably and interacting aberrantly with other cells. Apoptotic cell death is an early response of embryos to many if not all teratogens. Often, teratogen-induced cell death occurs in the areas of normal programmed cell death but in an expanded area. If cell death is not too extensive, embryos are thought to recover by compensatory cell proliferation. Excessive teratogen-induced cell death, however, is directly linked to abnormal development. For example, eye defects induced by 2-chloro-2'-deoxyadenosine are associated with excessive teratogen-induced cell death.

The intracellular signals of apoptosis are not yet known. Key components in the execution phase of the apoptotic pathway are the intracellular cysteinyl-aspartate proteases known as caspases, particularly caspase-3. These enzymes are normally present in all cells as inactive precursors that become activated by cleavage at specific internal motifs, in response to cytochrome c leaked by mitochondria into the cell's cytoplasm. Once activated, these caspases function to degrade specific target substrates such as poly(ADP-ribose)-polymerase (PARP), DNA-PKs, and lamins. Thereafter, chromosomal DNA is broken down. Treatment of cells with such developmental toxicants as hyperthermia, cyclophosphamide (an alkylating agent), and sodium arsenite (thiol oxidant) leads to the activation of caspase-3, cleavage of PARP, fragmentation of DNA, and cell death. It is not known how cells in the embryo recognize exposure to a developmental toxicant and initiate the apoptotic response, but perturbation of the redox status of the cell and oxidative stress are often, if not always, involved. As in other pathways, the apoptotic pathway engages in cross-talk, for example, with the nuclear factor-kappaB (NF-kB) and INF and FAS pathways. A recent report demonstrates that heat shock (43 °C) can rapidly activate the stress-activated protein kinase pathways mediated by c-JUN terminal kinase (JNK) and p38.

These molecular-stress and checkpoint pathways deserve attention as elements of the organism's defense against physical and chemical interventions. It remains to be learned whether polymorphisms of defense components exist in humans, compromising their responses to environmental agents. The extent to which the germ line, gametes, and early embryos operate these molecular-stress and checkpoint pathways is also poorly understood.

Developmental Differences

Although Drosophila and mouse development share more similarities than anyone thought 15 years ago, significant differences do exist. Mice share more aspects of development with other chordates (the chordate phylum includes vertebrates, cephalochordates, and urochordates) than they do with Drosophila, and they share still more aspects with other mammals. There appear to be "nested similarities" of development (i.e., the more recent the common ancestor of two groups, the more shared features of their development). Regarding HOX genes, for example, chordates have four more kinds of genes (HOX_{10-13}) than do arthropods. These differ slightly in sequence from the others and are located at the 5′ end of each cluster. They are expressed in the postanal tail, which is a chordate structure not shared by arthropods, and also in the developing vertebrate limb. Still, the difference between chordates and arthropods is a modification of a shared feature, namely, the use of HOX genes to divide the anteroposterior dimension of the animal into non-equivalent spatial compartments.

Chordates, but not arthropods, share the development of a dorsal hollow nerve cord, a notochord, and a segmentally arranged pharyngo-branchial apparatus, in addition to a postanal tail. They also share a kind of development involving a centralized "organizer" group of cells, the Spemann organizer, which releases inducers important in the placement, orientation, and scaling of later development by surrounding cells. The inducers secreted by the organizer have now been identified.

Several inducers are secreted protein antagonists of the TGFβ and WNT signals and are used by surrounding cells to maintain their ventral posterior paths of development. The inducer antagonists disinhibit and hence release the inherent capacity of the surrounding cells to undertake dorsal anterior kinds of development (e.g., to form the neural tube rather than epidermis). Few researchers would have guessed a few years ago that this subtle depression of neural development is the organizer's function in all chordates. Nonetheless, it should be noted that Holtfreter, building on a discovery of Barth, suggested that neural inducers provide little information except to release the inherent capacity of ectoderm cells to develop as neural tissue. This suggestion came from Barth and Holtfreter's findings that ectoderm would develop neural tissue if merely shocked briefly by ion imbalances or pH extremes.

Even though the organizer mode of development is distinctive to chordates, the components of the process are common to a wide range of other animals. For example, one antagonist, the Chordin protein, exhibits significant homology with the SOG protein of Drosophila. The SOG protein antagonizes a TGFβ inductive signal (called Screw) in Drosophila as part of the development of regions of neural versus epidermal development. Furthermore, in both Drosophila and frogs, there is a specific metalloproteinase that degrades the signal-antagonist complex, releasing the signal. The chordate and Drosophila inductive processes have deep similarities, though differing in details of time, place, and circumstances of use.

As a final example of differences, the dorsoventral dimension of arthropods looks quite different from that of a mouse, but recent analysis has shown that a number of similar genes are expressed in the nerve cords, hearts, body muscle, visceral mesoderm, and gut of both. It is currently accepted that these organs were present in primordial form in a common ancestor, but the arrangement of the organs in chordates is the inverse of that in arthropods. That is, the nerve cord is dorsal in chordates and ventral in arthropods, and the heart is ventral in chordates and dorsal in arthropods. The inversion of the dorsoventral axis is thought to have occurred in the chordate line after hemichordates split off.

Recognizing the fact that Drosophila does not share all details of early development and organogenesis with vertebrates, researchers have begun a systematic collection of developmental mutants of the zebrafish, a small vertebrate with a short life cycle, suitable for the production of a large mutant collection. The organs of embryonic zebrafish, more than the organs of Drosophila, resemble those of mammalian embryos in structure and function. In light of the extensive conservation of developmental processes found thus far, it is expected that in most cases what is true for fish development, as learned from those mutants, will be true for mammalian development, down to the level of molecular details of components and processes. That is not meant to deny differences among organisms (e.g., mammals undergo placental development with extensive extra-embryonic tissues not found in a zebrafish), nor to dismiss the possibility that developmental biologists might be misled in some instances by the study of model organisms. The greater part of mammalian development can be understood, however, by the study of other organisms' development. Ultimately, mammalian development will have to be understood in all the details of its differences, but even this pursuit will benefit from the context of knowledge of the processes shared with other organisms. For example, unique mammalian processes, such as extra-embryonic tissue formation or more extensive forebrain development, are still expected to entail many of the same signal transduction pathways and genetic regulatory circuits as used elsewhere in development.

The Evolutionary Perspective

In light of the availability of base sequences for a variety of kinds of genes in a variety of organisms, the place of metazoa (the multicellular animals) among the kingdoms of living organisms has been recently re-evaluated. It now appears that animals share a common ancestor more closely with plants (especially fungi) than with protozoa such as ciliates or amoebae. These three multicellular kingdoms arose from a common eukaryotic ancestor (probably single celled) perhaps 1.2 billion years ago, whereas eukaryotic single cells go back 2.2 to 2.7 billion years and prokaryotic life goes back perhaps 3.5 billion years. The conservation of basic biochemical, genetic, and cell biological functions has been surprisingly extensive in that long lineage. At least 3 billion years ago, ancient prokaryotes originated the processes of replication, transcription, translation, energy metabolism, and biosynthesis, and those processes have been carried forward to this day with little change in all life forms, including animals. The comparisons of the whole genomes of bacteria, yeast, and now the nematode, C. elegans, show clearly the conservation of the protein-coding sequences of genes. At least 2 billion years ago, single-celled eukaryotes originated the basic cell biological processes of mitosis, meiosis, a cdk-cyclin-based cell cycle, an actin-based cytoskeleton and myosin-based movements, a tubulin-based cytoskeleton and kinesin-dynein-based movements, membrane-trafficking, and membrane-bounded organelles. These processes and structures have been carried forward by the single-celled eukaryotes and animals with little change to this day.

In light of this conservation of ancient processes, what have metazoa added in the past 1.2 billion years? Their innovations include abundant cell-cell signaling, extracellular matrix, cell junctions, and a wide range of responses to intercellular signals based on complex genetic regulatory circuits and protein phosphorylation. The C. elegans genome shows that, compared with yeast, metazoa have greatly expanded the number of genes encoding proteins of signal transduction, the cytoskeleton, and transcriptional regulation and have greatly increased the size and complexity of the cis-regulatory regions of genes. Metazoa seem to have evolved in a regulatory or informational direction, that of determining the time, location, and circumstances within a multicellular population for activating and inhibiting the many conserved biochemical and cell biological processes brought forward from their single-celled ancestors. All these ancient processes have been made contingent on cell-cell signaling.

Gametogenesis

The development and production of the male and female germ cells required to form a new individual.

The male and female germ cells are called gametes. The gametes in human males are produced by the testes, two globe-shaped reproductive organs just below the penis. Male gametes are what most people refer to as sperm. Gametes in human females are produced by the ovaries, two oblong organs on each side of the uterus in the lower abdomen. Female gametes are what most people refer to as eggs or ova. After sexual intercourse, an ejaculated sperm cell penetrates an egg and unites with it (fertilizes it). The fertilized egg is called the zygote.

The reproductive organs in both males and females (testes and ovaries, respectively) begin gametogenesis with a primitive germ cell. A primitive germ cell is a seed cell. Like a seed planted in a garden, a primitive germ cell initiates the process that eventually results in a new being. The primitive germ cell contains 46 pairs of chromosomes. Chromosomes are structures that hold the genetic information (the DNA) that determine the makeup of the new being. In humans, chromosomes influence hair, eye and skin color, height, bone structure, and all of the characteristics that prompt people to say that a child "takes after" his mother or father.

Chromosomes occur in pairs because they reflect the makeup of the previous generation 23 chromosomes from the father and 23 from the mother. However, the gametes produced by the testes and the ovaries cannot each contain 46 chromosomes. Otherwise, after they unite, they will contain 92 chromosomes. Thus, the germ cells produced by the testes and ovaries each divide once, then divide again, in a reduction process that creates cells containing 23 chromosomes, or half the original number. This reduction process is known as meiosis.

Then, after a male ejaculates sperm into a female and fertilizes her egg, a new individual with 46 chromosomes begins to form. This is the beginning of pregnancy.

"Gamete" is derived from the Greek word "gamete" (wife) and "gamein" (to marry). "Genesis" is derived from the Greek word "genein" (to produce). Thus, in gametogenesis, cells marry ("gamein") and produce ("genein") a new being.

Spermatogenesis

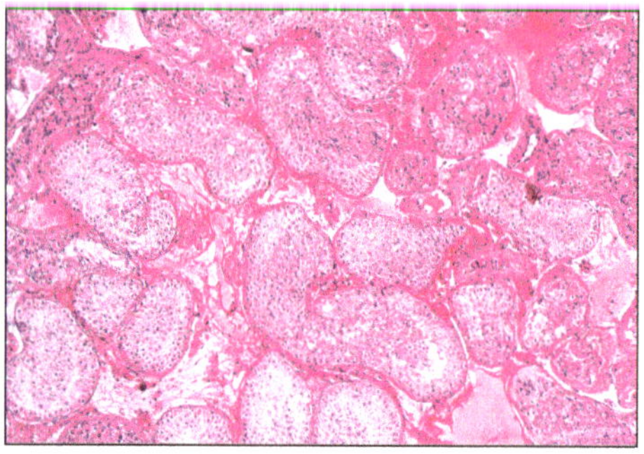

Normal spermatogenesis, testis biopsy.

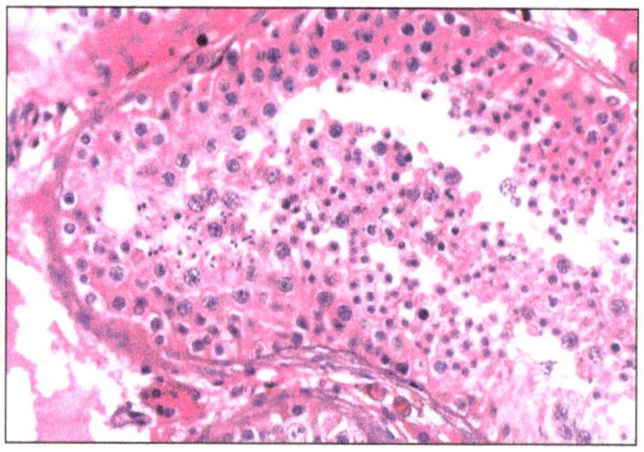

High-power view of a seminiferous tubule with normal spermatogenesis.

Spermatogenesis is the process by which haploid spermatozoa develop from germ cells in the seminiferous tubules of the testis. This process starts with the mitotic division of the stem cells located close to the basement membrane of the tubules. These cells are called spermatogonial stem cells. The mitotic division of these produces two types of cells. Type A cells replenish the stem cells, and type B cells differentiate into spermatocytes. The primary spermatocyte divides meiotically (Meiosis I) into two secondary spermatocytes; each secondary spermatocytes divides into two equal haploid spermatids by Meiosis II. The spermatids are transformed into spermatozoa(sperm) by the process of spermiogenesis. These develop into mature spermatozoa, also known as sperm cells. Thus, the primary spermatocyte gives rise to two cells, the secondary spermatocytes, and the two secondary spermatocytes by their subdivision produce four spermatozoa and four haploid cells.

Spermatozoa are the mature male gametes in many sexually reproducing organisms. Thus, spermatogenesis is the male version of gametogenesis, of which the female equivalent is oogenesis. In mammals it occurs in the seminiferous tubules of the male testes in a stepwise fashion. Spermatogenesis is highly dependent upon optimal conditions for the process to occur correctly, and is essential for sexual reproduction. DNA methylation and histone modification have been implicated in the regulation of this process. It starts at puberty and usually continues uninterrupted until death, although a slight decrease can be discerned in the quantity of produced sperm with increase in age.

Purpose

Spermatogenesis produces mature male gametes, commonly called *sperm* but more specifically known as *spermatozoa*, which are able to fertilize the counterpart female gamete, the oocyte, during conception to produce a single-celled individual known as a zygote. This is the cornerstone of sexual reproduction and involves the two gametes both contributing half the normal set of chromosomes (haploid) to result in a chromosomally normal (diploid) zygote.

To preserve the number of chromosomes in the offspring – which differs between species – one of each gamete must have half the usual number of chromosomes present in other body cells. Otherwise, the offspring will have twice the normal number of chromosomes, and serious abnormalities may result. In humans, chromosomal abnormalities arising from incorrect spermatogenesis results in congenital defects and abnormal birth defects (Down syndrome, Klinefelter syndrome) and in most cases, spontaneous abortion of the developing foetus.

Location in Humans

Spermatogenesis takes place within several structures of the male reproductive system. The initial stages occur within the testes and progress to the epididymis where the developing gametes mature and are stored until ejaculation. The seminiferous tubules of the testes are the starting point for the process, where spermatogonial stem cells adjacent to the inner tubule wall divide in a centripetal direction—beginning at the walls and proceeding into the innermost part, or *lumen*—to produce immature sperm. Maturation occurs in the epididymis. The location [Testes/Scrotum] is specifically important as the process of spermatogenesis requires a lower temperature to produce viable sperm, specifically 1°-8 °C lower than normal body temperature of 37 °C (98.6 °F). Clinically, small fluctuations in temperature such as from an athletic support strap, causes no impairment in sperm viability or count.

Duration

For humans, the entire process of spermatogenesis is variously estimated as taking 74 days (according to tritium-labelled biopsies) and approximately 120 days (according to DNA clock measurements). Including the transport on ductal system, it takes 3 months. Testes produce 200 to 300 million spermatozoa daily. However, only about half or 100 million of these become viable sperm.

Stages

The entire process of spermatogenesis can be broken up into several distinct stages, each corresponding to a particular type of cell in humans. In the following table, ploidy, copy number and chromosome/chromatid counts are for one cell, generally prior to DNA synthesis and division (in G1 if applicable). The primary spermatocyte is arrested after DNA synthesis and prior to division.

Cell type	ploidy/chromosomes in human	DNA copy number/chromatids in human	Process entered by cell
spermatogonium (types Ad, Ap and B)	diploid (2N) / 46	2C / 46	spermatocytogenesis (mitosis)
primary spermatocyte	diploid (2N) / 46	4C / 2x46	spermatidogenesis (meiosis I)
two secondary spermatocytes	haploid (N) / 23	2C / 2x23	spermatidogenesis (meiosis II)
four spermatids	haploid (N) / 23	C / 23	spermiogenesis
four functional spermatozoids	haploid (N) / 23	C / 23	spermiation

Spermatocytogenesis

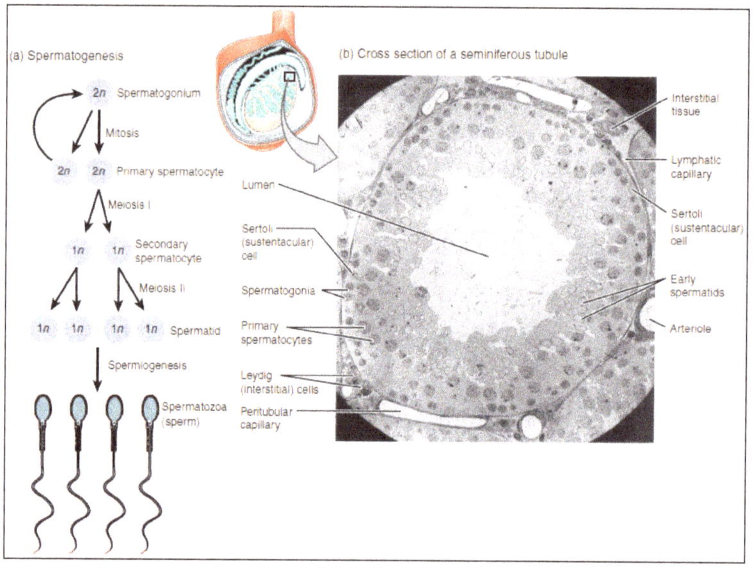

The process of spermatogenesis as the cells progress from primary spermatocytes, to secondary spermatocytes, to spermatids, to Sperm.

Spermatocytogenesis is the male form of gametocytogenesis and results in the formation of spermatocytes possessing half the normal complement of genetic material. In spermatocytogenesis, a diploid spermatogonium, which resides in the basal compartment of the seminiferous tubules, divides mitotically, producing two diploid intermediate cells called primary spermatocytes. Each primary spermatocyte then moves into the adluminal compartment of the seminiferous tubules and duplicates its DNA and subsequently undergoes *meiosis I* to produce two haploid secondary spermatocytes, which will later divide once more into haploid spermatids. This division implicates sources of genetic variation, such as random inclusion of either parental chromosomes, and chromosomal crossover that increases the genetic variability of the gamete. The DNA damage response (DDR) machinery plays an important role in spermatogenesis. The protein FMRP binds to meiotic chromosomes and regulates the dynamics of the DDR machinery during spermatogenesis. FMRP appears to be necessary for the repair of DNA damage.

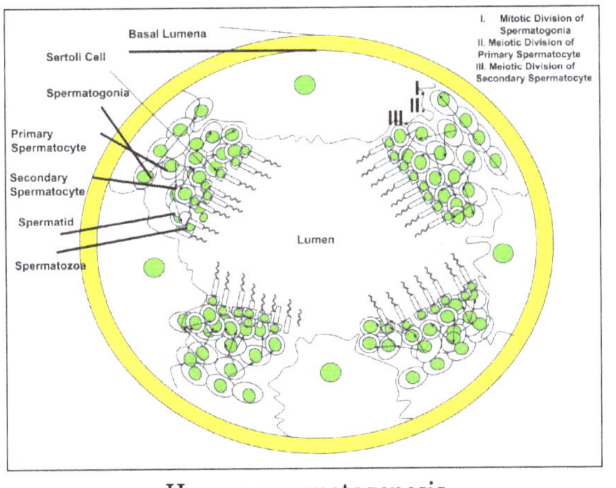

Human spermatogenesis

Each cell division from a spermatogonium to a spermatid is incomplete; the cells remain connected to one another by bridges of cytoplasm to allow synchronous development. It should also be noted that not all spermatogonia divide to produce spermatocytes; otherwise, the supply of spermatogonia would run out. Instead, spermatogonial stem cells divide mitotically to produce copies of themselves, ensuring a constant supply of spermatogonia to fuel spermatogenesis.

Spermatidogenesis

Spermatidogenesis is the creation of spermatids from secondary spermatocytes. Secondary spermatocytes produced earlier rapidly enter meiosis II and divide to produce haploid spermatids. The brevity of this stage means that secondary spermatocytes are rarely seen in histological studies.

Spermiogenesis

During spermiogenesis, the spermatids begin to form a tail by growing microtubules on one of the centrioles, which turns into basal body. These microtubules form an axoneme. Later the centriole is modified in the process of centrosome reduction. The anterior part of the tail (called midpiece) thickens because mitochondria are arranged around the axoneme to ensure energy supply. Spermatid DNA also undergoes packaging, becoming highly condensed. The DNA is packaged firstly with specific nuclear basic proteins, which are subsequently replaced with protamines during spermatid elongation. The resultant tightly packed chromatin is transcriptionally inactive. The Golgi apparatus surrounds the now condensed nucleus, becoming the acrosome.

Maturation then takes place under the influence of testosterone, which removes the remaining unnecessary cytoplasm and organelles. The excess cytoplasm, known as *residual bodies*, is phagocytosed by surrounding Sertoli cells in the testes. The resulting spermatozoa are now mature but lack motility, rendering them sterile. The mature spermatozoa are released from the protective Sertoli cells into the lumen of the seminiferous tubule in a process called *spermiation*.

The non-motile spermatozoa are transported to the epididymis in *testicular fluid* secreted by the Sertoli cells with the aid of peristaltic contraction. While in the epididymis the spermatozoa gain motility and become capable of fertilization. However, transport of the mature spermatozoa through the remainder of the male reproductive system is achieved via muscle contraction rather than the spermatozoon's recently acquired motility.

Role of Sertoli Cells

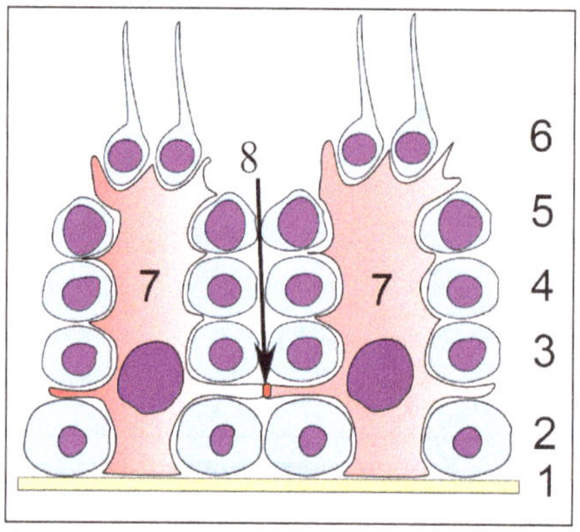

Labelled diagram of the organisation of Sertoli cells (red) and spermatocytes (blue) in the testis. Spermatids which have not yet undergone spermiation are attached to the lumenal apex of the cell.

At all stages of differentiation, the spermatogenic cells are in close contact with Sertoli cells which are thought to provide structural and metabolic support to the developing sperm cells. A single Sertoli cell extends from the basement membrane to the lumen of the seminiferous tubule, although the cytoplasmic processes are difficult to distinguish at the light microscopic level.

Sertoli cells serve a number of functions during spermatogenesis, they support the developing gametes in the following ways:

- Maintain the environment necessary for development and maturation, via the blood-testis barrier.
- Secrete substances initiating meiosis.
- Secrete supporting testicular fluid.
- Secrete androgen-binding protein (ABP), which concentrates testosterone in close proximity to the developing gametes:
    - Testosterone is needed in very high quantities for maintenance of the reproductive tract, and ABP allows a much higher level of fertility.
- Secrete hormones affecting pituitary gland control of spermatogenesis, particularly the polypeptide hormone, inhibin.
- Phagocytose residual cytoplasm left over from spermiogenesis.

- Secretion of anti-Müllerian hormone causes deterioration of the Müllerian duct.
- Protect spermatids from the immune system of the male, via the blood-testis barrier.
- Contribute to the spermatogonial stem cell niche.

The intercellular adhesion molecules ICAM-1 and soluble ICAM-1 have antagonistic effects on the tight junctions forming the blood-testis barrier. ICAM-2 molecules regulate spermatid adhesion on the apical side of the barrier (towards the lumen).

Influencing Factors

The process of spermatogenesis is highly sensitive to fluctuations in the environment, particularly hormones and temperature. Testosterone is required in large local concentrations to maintain the process, which is achieved via the binding of testosterone by androgen binding protein present in the seminiferous tubules. Testosterone is produced by interstitial cells, also known as Leydig cells, which reside adjacent to the seminiferous tubules.

Seminiferous epithelium is sensitive to elevated temperature in humans and some other species, and will be adversely affected by temperatures as high as normal body temperature. Consequently, the testes are located outside the body in a sack of skin called the scrotum. The optimal temperature is maintained at 2 °C (man)–8 °C (mouse) below body temperature. This is achieved by regulation of blood flow and positioning towards and away from the heat of the body by the cremasteric muscle and the dartos smooth muscle in the scrotum.

Dietary deficiencies (such as vitamins B, E and A), anabolic steroids, metals (cadmium and lead), x-ray exposure, dioxin, alcohol, and infectious diseases will also adversely affect the rate of spermatogenesis. In addition, the male germ line is susceptible to DNA damage caused by oxidative stress, and this damage likely has a significant impact on fertilization and pregnancy. Exposure to pesticides also affects spermatogenesis.

Hormonal Control

Hormonal control of spermatogenesis varies among species. In humans the mechanism is not completely understood; however it is known that initiation of spermatogenesis occurs at puberty due to the interaction of the hypothalamus, pituitary gland and Leydig cells. If the pituitary gland is removed, spermatogenesis can still be initiated by follicle stimulating hormone (FSH) and testosterone. In contrast to FSH, LH appears to have little role in spermatogenesis outside of inducing gonadal testosterone production.

FSH stimulates both the production of androgen binding protein (ABP) by Sertoli cells, and the formation of the blood-testis barrier. ABP is essential to concentrating testosterone in levels high enough to initiate and maintain spermatogenesis. Intratesticular testosterone levels are 20–100 or 50–200 times higher than the concentration found in blood, although there is variation over a 5- to 10-fold range amongst healthy men. FSH may initiate the sequestering of testosterone in the testes, but once developed only testosterone is required to maintain spermatogenesis. However, increasing the levels of FSH will increase the production of spermatozoa by preventing the apoptosis of *type A spermatogonia*. The hormone inhibin acts to decrease the levels of FSH. Studies from rodent models suggest that gonadotropins (both LH and FSH) support the process of spermatogenesis by suppressing the proapoptotic signals and therefore promote spermatogenic cell survival.

The Sertoli cells themselves mediate parts of spermatogenesis through hormone production. They are capable of producing the hormones estradiol and inhibin. The Leydig cells are also capable of producing estradiol in addition to their main product testosterone. Estrogen has been found to be essential for spermatogenesis in animals. However, a man with estrogen insensitivity syndrome (a defective ERα) was found produce sperm with a normal sperm count, albeit abnormally low sperm viability; whether he was sterile or not is unclear. Levels of estrogen that are too high can be detrimental to spermatogenesis due to suppression of gonadotropin secretion and by extension intratesticular testosterone production. Prolactin also appears to be important for spermatogenesis.

Oogenesis

Oogenesis—the differentiation of the ovum—differs from spermatogenesis in several ways. Whereas the gamete formed by spermatogenesis is essentially a motile nucleus, the gamete formed by oogenesis contains all the materials needed to initiate and maintain metabolism and development. Therefore, in addition to forming a haploid nucleus, oogenesis also builds up a store of cytoplasmic enzymes, mRNAs, organelles, and metabolic substrates. While the sperm becomes differentiated for motility, the egg develops a remarkably complex cytoplasm.

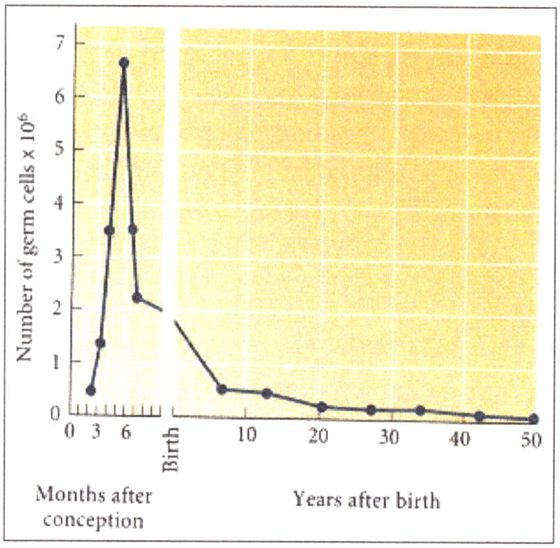

Changes in the number of germ cells in the human ovary over the life span.

The mechanisms of oogenesis vary among species more than those of spermatogenesis. This difference should not be surprising, since patterns of reproduction vary so greatly among species. In some species, such as sea urchins and frogs, the female routinely produces hundreds or thousands of eggs at a time, whereas in other species, such as humans and most mammals, only a few eggs are produced during the lifetime of an individual. In those species that produce thousands of ova, the oogonia are self-renewing stem cells that endure for the lifetime of the organism. In those species that produce fewer eggs, the oogonia divide to form a limited number of egg precursor cells. In the human embryo, the thousand or so oogonia divide rapidly from the second to the seventh month of gestation to form roughly 7 million germ cells .After the seventh month of embryonic development, however, the number of germ cells drops precipitously. Most oogonia die during this period, while the remaining oogonia enter the first meiotic division. These latter cells, called the primary oocytes, progress through the first meiotic prophase until the diplotene stage, at which point they are maintained until puberty. With the onset of adolescence, groups of oocytes periodically resume meiosis. Thus, in the human female, the first part of meiosis begins in the embryo, and the signal to resume meiosis is not given until roughly 12 years later. In fact, some oocytes are maintained in meiotic prophase for nearly 50 years. Primary oocytes continue to die even after birth. Of the millions of primary oocytes present at birth, only about 400 mature during a woman's lifetime.

Oogenic meiosis also differs from spermatogenic meiosis in its placement of the metaphase plate. When the primary oocyte divides, its nucleus, called the germinal vesicle, breaks down, and the metaphase spindle migrates to the periphery of the cell. At telophase, one of the two daughter cells contains hardly any cytoplasm, whereas the other cell has nearly the entire volume of cellular constituents The smaller cell is called the first polar body, and the larger cell is referred to as the secondary oocyte. During the second division of meiosis, a similar unequal cytokinesis takes place. Most of the cytoplasm is retained by the mature egg (ovum), and a second polar body receives little more than a haploid nucleus. Thus, oogenic meiosis conserves the volume of oocyte cytoplasm in a single cell rather than splitting it equally among four progeny.

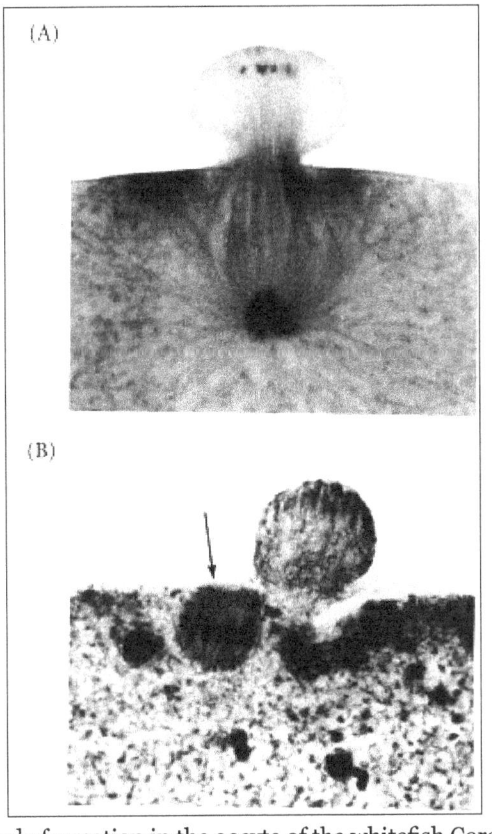

Polar body formation in the oocyte of the whitefish Coregonus.

In figure, (A) Anaphase of first meiotic division, showing the first polar body pinching off with its chromosomes. (B) Metaphase (within the oocyte, arrow) of the second meiotic division, with the first polar body still in place. The first polar body may or may not divide again.

In a few species of animals, meiosis is severely modified such that the resulting gamete is diploid and need not be fertilized to develop. Such animals are said to be parthenogenetic (Greek, "virgin birth"). In the fly Drosophila mangabeirai, one of the polar bodies acts as a sperm and "fertilizes" the oocyte after the second meiotic division. In other insects (such as Moraba virgo) and in the lizard Cnemidophorus uniparens, the oogonia double their chromosome number before meiosis, so that the halving of the chromosomes restores the diploid number. The germ cells of the grasshopper Pycnoscelus surinamensis dispense with meiosis altogether, forming diploid ova by two mitotic divisions. All of these species consist entirely of females. In other species, haploid parthenogenesis is widely used not only as a means of reproduction, but also as a mechanism of sex determination. In the Hymenoptera (bees, wasps, and ants), unfertilized eggs develop into males, whereas fertilized eggs, being diploid, develop into females. The haploid males are able to produce sperm by abandoning the first meiotic division, thereby forming two sperm cells through second meiosis.

Maturation of the Oocyte in Amphibians

The egg is responsible for initiating and directing development, and in some species (as seen above), fertilization is not even necessary. The accumulated material in the oocyte cytoplasm includes energy sources and energy-producing organelles (the yolk and mitochondria); the enzymes and precursors for DNA, RNA, and protein syntheses; stored messenger RNAs; structural proteins; and morphogenetic regulatory factors that control early embryogenesis. A partial catalogue of the materials stored in the oocyte cytoplasm is shown in table, while a partial list of stored mRNAs is shown in table. Most of this accumulation takes place during meiotic prophase I, and this stage is often subdivided into two phases, previtellogenesis and vitellogenesis.

Table: Cellular components stored in the mature oocyte of Xenopus laevis.

Component	Approximate excess over amount in larval cells
Mitochondria	100,000
RNA polymerases	60,000–100,000
DNA polymerases	100,000
Ribosomes	200,000
tRNA	10,000
Histones	15,000
Deoxyribonucleoside triphosphates	2,500

The eggs of fishes and amphibians are derived from an oogonial stem cell population that can generate a new cohort of oocytes each year. In the frog Rana pipiens, oogenesis takes 3 years. During the first 2 years, the oocyte increases its size very gradually. During the third year, however, the rapid accumulation of yolk in the oocyte causes the egg to swell to its characteristically large size. Eggs mature in yearly batches, with the first cohort maturing shortly after metamorphosis; the next group matures a year later.

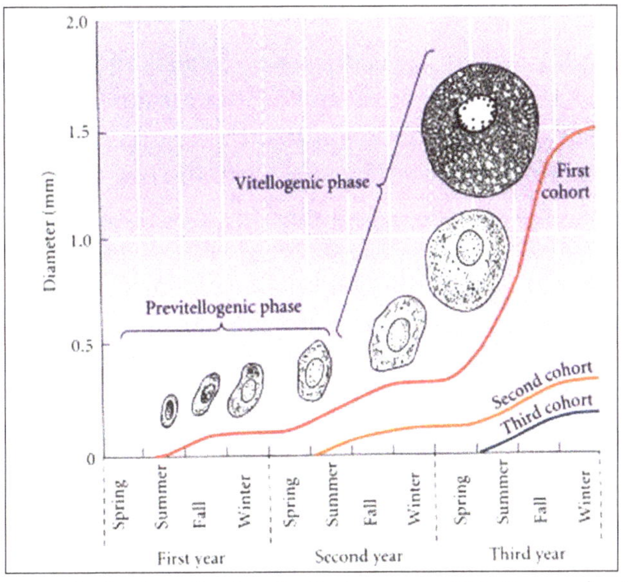

Growth of oocytes in the frog. During the first 3 years of life, three cohorts of oocytes are produced. The drawings follow the growth of the first-generation oocytes.

Vitellogenesis occurs when the oocyte reaches the diplotene stage of meiotic prophase. Yolk is not a single substance, but a mixture of materials used for embryonic nutrition. The major yolk component in frog eggs is a 470-kDa protein called vitellogenin. It is not made in the frog oocyte (as are the major yolk proteins of organisms such as annelids and crayfishes), but is synthesized in the liver and carried by the bloodstream to the ovary. This large protein passes between the follicle cells of the ovary, and is incorporated into the oocyte by micropinocytosis, the pinching off of membrane-bounded vesicles at the bases of microvilli. In the mature oocyte, vitellogen-in is split into two smaller proteins: the heavily phosphorylatedphosvitin and the lipoprotein lipovitellin. These two proteins are packaged together into membrane-bounded yolk platelets Glycogen granules and lipochondrial inclusions store the carbohydrate and lipid components of the yolk, respectively.

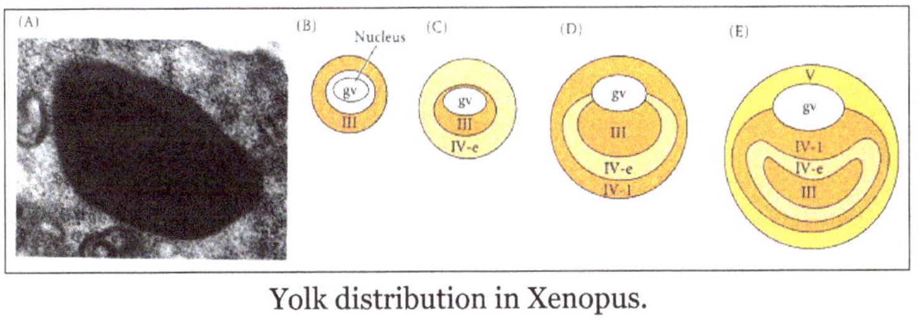

Yolk distribution in Xenopus.

In figure, (A) An amphibian yolk platelet. (B-E) Establishment of animal-vegetal distribution of yolk platelets in the Xenopus oocyte. (B) In the late stage III (600 μm) oocyte, yolk platelets enter the cell equally at all points of the surface. (C, D) As the oocyte grows, the platelets at the future animal pole are displaced toward the vegetal pole, while those at the vegetal pole remain there. More yolk still enters the egg on all sides. (E) By the end of vitellogenesis, the earliest platelets (III) are all in the vegetal hemisphere, which now contains roughly 75% of the oocyte yolk. Timing of entry of yolk into oocyte platelets is indicated by shading and Roman numerals: III, stage III platelets; IV-e, early stage IV platelets; IV-l, late stage IV platelets; V, stage V platelets; gv, germinal vesicle.

Most eggs are highly asymmetrical, and it is during oogenesis that the animal-vegetal axis of the egg is specified. Danilchik and Gerhart have shown that although the concentration of yolk increases nearly tenfold as one moves from the animal to the vegetal poles of the mature Xenopus egg, vitellogenin uptake is uniform around the surface of the oocyte. What varies is its movement within the oocyte, and this depends on where the yolk proteins enter. When yolk platelets are formed in the future animal hemisphere, they move inward toward the center of the cell. Vegetal yolk platelets, however, do not actively move, but remain at the periphery of the cell for long periods of time, enlarging as they stay there. They are slowly displaced from the cortex as new yolk platelets come in from the surface. As a result of this differential intracellular transport, the amount of yolk steadily increases in the vegetal hemisphere, until the vegetal half of a mature Xenopus oocyte contains nearly 75% of the yolk. The mechanism of this translocation remains unknown.

As the yolk is being deposited, the organelles also become arranged asymmetrically. The cortical granules begin to form from the Golgi apparatus; they are originally scattered randomly through the oocyte cytoplasm, but later migrate to the periphery of the cell. The mitochondria replicate at this time, dividing to form millions of mitochondria that will be apportioned to the different blastomeres during cleavage. (In Xenopus, new mitochondria will not be formed until after gastrulation is initiated.) As vitellogenesis nears an end, the oocyte cytoplasm becomes stratified. The cortical granules, mitochondria, and pigment granules are found at the periphery of the cell, within the actin-rich oocyte cortex. Within the inner cytoplasm, distinct gradients emerge. While the yolk platelets become more heavily concentrated at the vegetal pole of the oocyte, the glycogen granules, ribosomes, lipid vesicles, and endoplasmic reticulum are found toward the animal pole. Even specific mRNAs stored in the cytoplasm become localized to certain regions of the oocyte.

While the precise mechanisms for establishing these gradients remain unknown, studies using inhibitors have shown that the cytoskeleton is critically important in localizing specific RNAs and morphogenetic factors. There seem to be two pathways for gettting mRNAs into the vegetal cortex. The first pathway moves messages such as those encoding the Vg1 protein, which are initially present throughout the oocyte, into the vegetal cortex in a two-step process. In the first phase, microtubules are needed to bring Vg1 mRNA into the vegetal hemisphere. In the second phase, microfilaments are responsible for anchoring the Vg1 message to the cortex. The portion of the Vg1 mRNA that binds to these cytoskeletal elements resides in its 3′ untranslated region. When a specific 340-base sequence from the Vg1 3′ UTR is placed onto a β-globin message, that β-globin mRNA is similarly localized to the vegetal cortex. Other mRNAs, including germ plasm mRNAS such as Xlsirt and Xcat2, leave the germinal vesicle and join the mitochondrial "cloud" located at the vegetal pole of the nucleus. These messages are compartmentalized into clusters associated with the germ plasm and transported to the vegetal cortex in a manner that appears to be independent of the cytoskeleton. This mechanism is known as the Metro (message transport organizer) pathway.

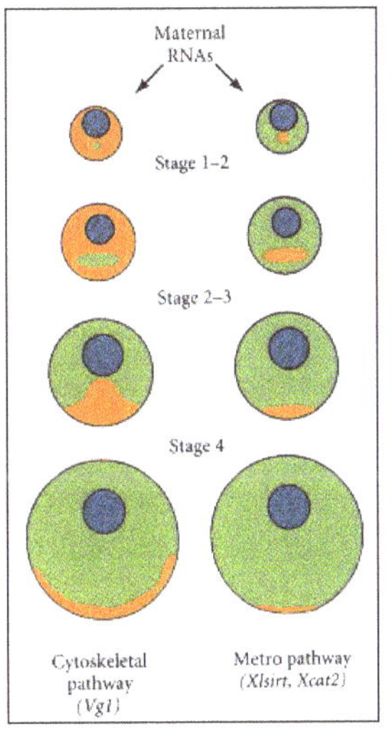

Schematic representations of the two pathways for localizing mRNAs to the vegetal region of the Xenopus oocyte. The respective mRNAs are shown in yellow. In the cytoskeletal (Vg1) pathway, messages are first seen throughout the egg, but they are translocated through a microtubule-driven system to the microfilaments of the vegetal cortex. The Metro (message transport organizer) pathway accumulates messages in the mitochondrial cloud, and islands of them are transported into the vegetal cortex.

Completion of Amphibian Meiosis: Progesterone and Fertilization

Amphibian oocytes can remain for years in the diplotene stage of meiotic prophase. This state resembles the G_2 phase of the cell division cycle. Resumption of meiosis in the amphibian primary oocyte requires progesterone. This hormone is secreted by the follicle cells in response to gonadotropic hormones secreted by the pituitary gland. Within 6 hours of progesterone stimulation, germinal vesicle breakdown (GVBD) occurs, the microvilli retract, the nucleoli disintegrate, and the chromosomes contract and migrate to the animal pole to begin division. Soon afterward, the first meiotic division occurs, and the mature ovum is released from the ovary by a process called ovulation. The ovulated egg is in second meiotic metaphase when it is released.

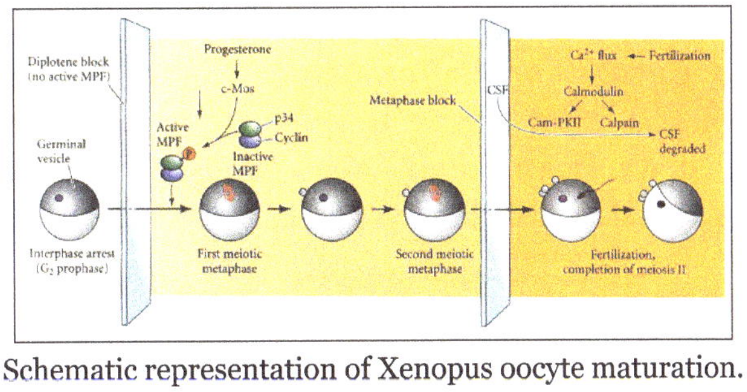

Schematic representation of Xenopus oocyte maturation.

In figure showing the regulation of meiotic cell division by progesterone and fertilization. Oocyte maturation is blocked in the diplotene stage of first meiotic prophase by the lack of active MPF. Progesterone activates the production of c-mos, and the c-mos protein initiates a cascade of phosphorylation that eventually phosphorylates the p34 component of MPF. This allows the MPF to become active. The MPF drives the cell cycle through the first meiotic division, but further division is blocked by CSF, a compound containing c-mos and cdk2. Upon fertilization, calcium ions released into the cytoplasm are bound by calmodulin and are used to activate two enzymes, calmodulin-dependent protein kinase II and calpain II, which inactivate and degrade CSF. Second meiosis is completed, and the two haploid pronuclei can fuse. At this time, cyclin B is resynthesized, allowing first cell cycle of cleavage to begin.

How does progesterone enable the egg to break its dormancy and resume meiosis? To understand the mechanisms by which this activation is accomplished, it is necessary to briefly review the model for early blastomere division. Entry into the mitotic (M) phase of the cell cycle (in both meiosis and mitosis) is regulated by mitosis-promoting factor, or MPF (originally called "maturation-promoting factor" after its meiotic function). MPF contains two subunits, cyclin B and the $p34^{cdc2}$ protein. The p34 protein is a cyclin-dependent-kinase—its activity is dependent upon the presence of cyclin. Since all the components of MPF are present in the amphibian oocyte, it is generally thought that progesterone somehow converts a pre-MPF complex into active MPF.

The mediator of the progesterone signal is the c-mos protein. Progesterone reinitiates meiosis by causing the egg to polyadenylate the maternal c-mos mRNA that has been stored in its cytoplasm. This message is translated into a 39-kDa phosphoprotein, known as c-mos. This protein is detectable only during oocyte maturation and is destroyed quickly upon fertilization. Yet during its brief lifetime, it plays a major role in releasing the egg from its dormancy. If the translation of c-mos is inhibited (by injecting c-mos antisense mRNA into the oocyte), germinal vesicle breakdown and the resumption of oocyte maturation do not occur. The c-mos protein activates a phosphorylation cascade that phosphorylates and activates the p34 subunit of MPF. The active MPF allows the germinal vesicle to break down and the chromosomes to divide.

However, the chromosomes then encounter a second block. MPF can take the chromosomes through only the first meiotic division and the prophase of the second meiotic division. The oocyte is arrested again in the metaphase of the second meiotic division. This metaphase block is caused by the combined actions of c-mos and another protein, cyclin-dependent kinase. These two proteins are subunits of cytostatic factor (CSF), which is found in mature frog eggs, and which can block cell cycles in metaphase. It is thought that CSF prevents the degradation of cyclin.

The metaphase block is broken by fertilization. Evidence suggests that the calcium ion flux attending fertilization enables the calcium-binding protein calmodulin to become active. Calmodulin, in turn, can activate two enzymes that inactivate CSF: calmodulin-dependent protein kinase II, which inactivates p34, and calpain II, a calcium-dependent protease that degrades c-mos. Without CSF, cyclin can be degraded, and the meiotic division can be completed.

Gene Transcription in Oocytes

In most animals (insects being a major exception), the growing oocyte is active in transcribing genes whose products are (1) necessary for cell metabolism, (2) necessary for oocyte-specific processes, or (3) needed for early development before the zygote-derived nuclei begins to function. In mice, for instance, the growing diplotene oocyte is actively transcribing the genes for zona pellucida proteins ZP1, ZP2, and ZP3. Moreover, these genes are transcribed only in the oocyte and not in any other cell.

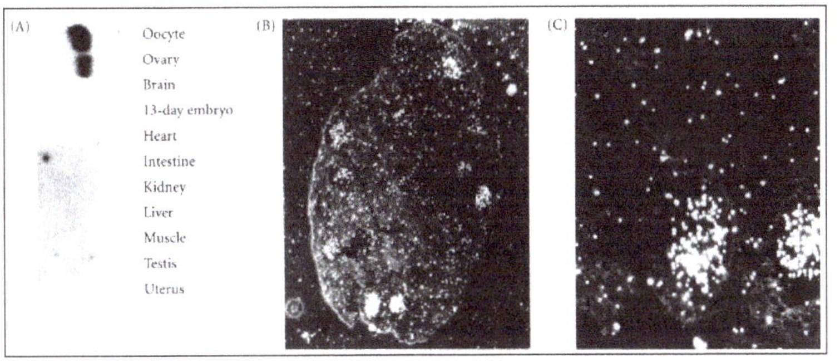

Expression of the ZP3 gene in the developing mouse oocyte.

In figure, (A) Northern blot of ZP3 mRNA accumulation in embryonic mouse tissues. A radioactive probe to the ZP3 message found it expressed only in the ovary, and specifically in the oocytes. (B-C) When the luciferinase reporter gene is placed onto the ZP3 promoter and inserted into the mouse genome, the luciferinase message is seen only in the developing oocytes of the ovary. C is a higher magnification of a section of B, showing two of the ovarian follicles containing maturing oocytes.

The amphibian oocyte has certain periods of very active RNA synthesis. During the diplotene stage, certain chromosomes stretch out large loops of DNA, causing the chromosome to resemble a lampbrush (a handy instrument for cleaning test tubes in the days before microfuges). These lamp brush chromosomes can be revealed as the sites of RNA synthesis by in situ hybridization. Oocyte chromosomes can be incubated with a radioactive RNA probe, and autoradiography used to visualize the precise location where the gene is being transcribed. Figure shows diplotene chromosome I of the newt Triturus cristatus after incubation with radioactive histone mRNA. It is obvious that a histone gene (or set of histone genes) is located on one of these loops of the lampbrush chromosome. Electron micrographs of gene transcripts from lampbrush chromosomes also enable one to see chains of mRNA coming off each gene as it is transcribed.

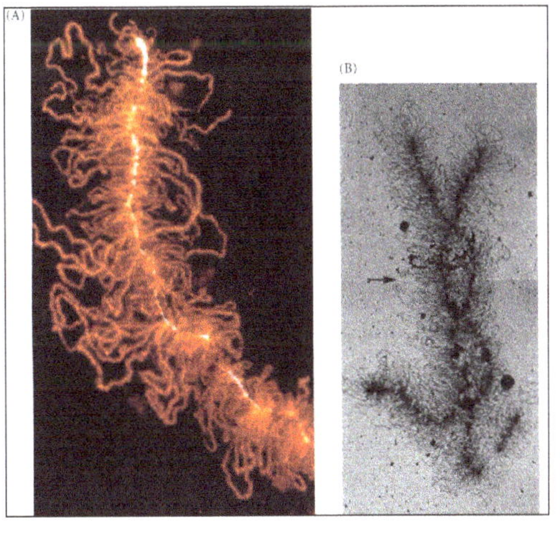

In figure shows amphibian lampbrush chromosomes are active in the diplotene germinal vesicle during first meiotic prophase. (A) A lampbrush chromosome of the salamander Notophthalmus viridescens. Extended DNA (white) loops out and is transcribed into RNA (red). (B) Localization (arrow) of histone genes on a lampbrush chromosome in an amphibian oocyte. The genes have been visualized by in situ hybridization and autoradiography.

In addition to mRNA synthesis, the patterns of ribosomal RNA and transfer RNA transcription are also regulated during oogenesis. Figure shows the course of rRNA and tRNA synthesis during Xenopus oogenesis. Transcription appears to begin in early (stage I, 25–40 μm) oocytes, during the diplotene stage of meiosis. At this time, all the rRNAs and tRNAs needed for protein synthesis until the mid-blastula stage are made, and all the maternal mRNAs needed for early development are transcribed. This stage lasts for months in Xenopus. The rate of ribosomal RNA production is prodigious. The Xenopus oocyte genome has over 1800 genes encoding 18S and 28S rRNA (the two large RNAs that form the ribosomes), and these genes are selectively amplified such that there are over 500,000 genes making these ribosomal RNAs. After reaching a certain size, the chromosomes of the mature (stage VI) oocyte condense, and the genes are no longer transcribed. This "mature oocyte" condition can also last for months. Upon hormonal stimulation, the oocyte completes its first meiotic division and is ovulated. The mRNAs stored by the oocyte now join with the ribosomes to initiate protein synthesis. Within hours, the second meiotic division has begun, and the secondary oocyte has been fertilized. The embryo's genes do not begin active transcription until the mid-blastula transition.

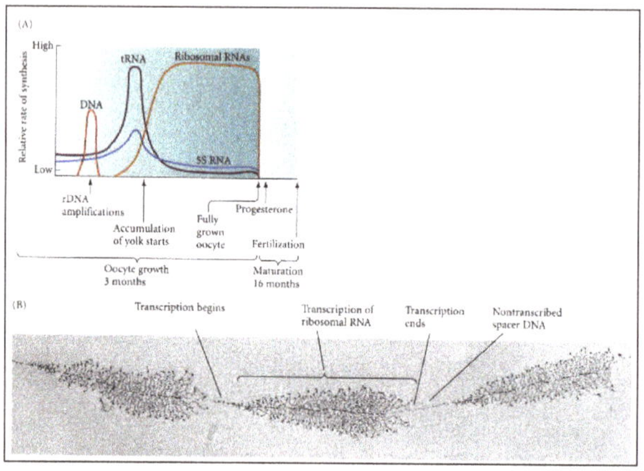

Ribosomal RNA production in Xenopus oocytes.

In figure, (A) Relative rates of DNA, tRNA, and rRNA synthesis in amphibian oogenesis during the last 3 months before ovulation. (B) The transcription of the RNA precursor of the 28S, 18S, and 5.8S ribosomal RNAs. These units are tandemly linked together, some 450 per haploid genome.

The oocytes of several species make two classes of mRNAs—those for immediate use in the oocyte and those that are stored for use during early development. In sea urchins, the translation of stored maternal messages is initiated by fertilization, while in frogs, the signal for such translation is initiated by progesterone as the egg is about to be ovulated. One of the results of the MPF activity induced by progesterone may be the phosphorylation of proteins on the 3′ UTR of stored oocyte mRNAs. The phosphorylation of these factors is associated with the lengthening of the poly(A) tails in the stored messages and the translation of the stored mRNAs.

Meroistic Oogenesis in Insects

There are several types of oogenesis in insects, but most studies have focused on those insects, such as Drosophila and moths, that undergo meroistic oogenesis, in which cytoplasmic connections remain between the cells produced by the oogonium. In Drosophila, each oogonium divides four times to produce a clone of 16 cells connected to each other by ring canals. The production of these interconnected cells, called cystocytes, involves a highly ordered array of cell divisions. Only those two cells having four interconnections are capable of developing into oocytes, and of those two, only one becomes the egg. The other begins meiosis but does not complete it. Thus, only one of the 16 cystocytes can become an ovum. All the other cells become nurse cells. As it turns out, the cell destined to become the oocyte is that cell residing at the most posterior tip of the egg chamber, or ovariole, that encloses the 16-cell clone. However, since the nurse cells are connected to the oocyte by cytoplasmic bridges, the entire complex can be seen as one egg-producing unit.

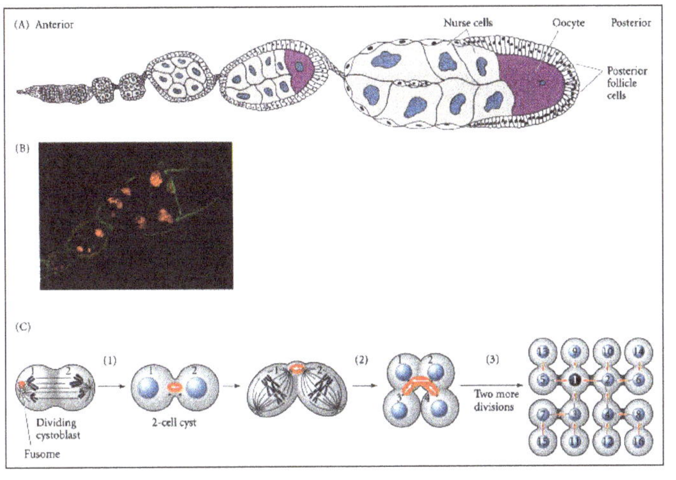

In figure show the formation of 16 interconnected cystocytes in Drosophila. (A) Diagram of adult ovariole, showing sequence of oogenesis as younger germinal cysts mature within the ovariole. (B) Section through center of an ovariole, showing the maturation of the oocyte. Actin is stained green, and the nuclear DNA is stained red. (C) Division of the cystocyte-forming cells (cystoblasts). The cells are represented schematically as dividing in a single plane. The stem cell divides to produce another stem cell plus an oogonium that is committed to forming the cystocytes. When the oogonium divides, the centriole of daughter cystocyte 1 retains the fusome (red), which grows through the ring canal toward its mitotic sister. The arrow shows the polarity, pointing to the cell from which the fusome grew. After three more mitotic divisions, the 16-cell cyst is formed. If intracellular transport is coordinated by the fusome, the transport of mRNAs and proteins would be toward cystocyte 1, which would thus become the oocyte.

The structure of the Meroistic Ovary

The meroistic ovary confronts us with some interesting problems. If all the cystocytes are connected so that proteins and RNAs shuttle freely among them, why should they have different developmental fates? Why should one cell become the oocyte while the others become "RNA-synthesizing factories," sending mRNAs, ribosomes, and even centrioles into the oocyte? Why is the flow of protein and RNA in one direction only? As the cystocytes divide, a large, spectrin-rich structure called the fusome forms and spans the ring canals between the cells. It is constructed asymmetrically, as it always grows from the spindle pole that remained in one of the cells after the first division. The cell that retained the greater part of the fusome during the first division becomes the oocyte. It is not yet known if the fusome contains oogenic determinants, or if it directs the traffic of materials into this particular cell.

Once the patterns of transport are established, the cytoskeleton becomes actively involved in transporting mRNAs from the nurse cells into the oocyte cytoplasm. The microtubular array is critical for oocyte determination. If this lattice is disrupted (either chemically or by mutations such as bicaudal-D or egalitarian), gene products are transmitted in all directions, and all 16 cells differentiate into nurse cells. It is possible that some compounds transported from the nurse cells into the oocyte become associated with transport proteins, such as kinesin, that would enable them to travel along the tracks of microtubules extending through the ring canals. Actin may become important for maintaining this polarity during later stages of oogenesis. Mutations that prevent actin microfilaments from lining the ring canals prevent the transport of mRNAs from nurse cells to oocyte, and disruption of the actin microfilaments randomizes the distribution of mRNA. Thus, the cytoskeleton appears to control the movement of organelles and RNAs between nurse cells and oocyte such that developmental cues are exchanged only in the appropriate direction.

Transport of RNA from Nurse Cells to Oocyte

The oocytes of meroistic insects do not pass through a transcriptionally active stage, nor do they have lampbrush chromosomes. Rather, RNA synthesis is largely confined to the nurse cells, and the RNA made by those cells is actively transported into the oocyte cytoplasm. Oogenesis takes place in only 12 days, so the nurse cells are very metabolically active during this time. They are aided in their transcriptional efficiency by becoming polytene. Instead of having two copies of each chromosome, they replicate their chromosomes until they have produced 512 copies. The 15 nurse cells pass ribosomal and messenger RNAs as well as proteins into the oocyte cytoplasm, and entire ribosomes may be transported as well. The mRNAs do not associate with polysomes, which suggests that they are not immediately active in protein synthesis.

Fertilization

Fertilization is a complicated multistep molecular process where two highly methylated and specified haploid gametes, spermatozoon and oocyte, are coming together forming a male and a female pronucleus, respectively, and culminating in the fusion of the two pronuclei giving rise to the formation of the zygote. It involves distinct events with the following order:

1. The migration of spermatozoa through the cumulus matrix,
2. The adhesion and penetration of spermatozoa to the zona pellucida,
3. The fusion of the two plasma membranes. Different molecules through complicated molecular pathways are participating in each step, and therefore, the functionality of each step depends upon the integrity of these molecules. Upon fusion of the plasma membranes,
4. Oocyte activation and a fast block to polyspermy events are taking place,
5. While the male and the female pronuclei are coming together forming the nucleus of the zygote and finalizing in this way the fertilization process.

Before we analyze the various phases of fertilization, it is worth mentioning that the spermatozoa after ejaculation are migrating from the uterus to the oviduct and arrive at the fertilization site in the ampulla. In this migration process, molecules, such as fertilin, have been implicated, while the spermatozoa "locate" the ovulated oocyte by chemoattractants released by the oocytes themselves.

Upon arrival to the fertilization site, a cascade of molecular and biochemical responses are initiated to enable fertilization to occur.

Sperm are produced by the testis and become mature and motile as they travel through the epididymis. Sperm released on ejaculation have to travel through the cervical os, the uterus, and the fallopian tubes where hopefully an oocyte is waiting for fertilization. The mature sperm consists of a head piece, neck, and flagellum. The flagellum is responsible for initiation and maintenance of motility through the female reproductive tract. The head of the sperm contains the sperm DNA and is the area involved in recognition of the zona pellucida (the glycoprotein coat that surrounds the oocyte) and sperm–oocyte fusion. The head is covered by a membrane-bound vesicle called the acrosome. Before a sperm can fertilize an oocyte it needs to go through capacitation and the acrosome reaction. Capacitation occurs in the female reproductive tract and involves a range of poorly understood processes that do not alter the ultrastructure of the sperm, but enable the sperm to fertilize the oocyte.

References

- Biology-developmental: plato.stanford.edu, Retrieved 1 January, 2019
- Developmental-biology-changing-lives: openaccessgovernment.org, Retrieved 3 February, 2019
- Developmental-biology-important: openaccessgovernment.org, Retrieved 5 March, 2019
- Ettensohn C.A., Sweet H.C. (2000). Patterning the early sea urchin embryo. Curr. Top. Dev. Biol. Current Topics in Developmental Biology. 50. pp. 1–44. doi:10.1016/S0070-2153(00)50002-7. ISBN 9780121531508. PMID 10948448
- Fertilization, medicine-and-dentistry: sciencedirect.com, Retrieved 7 April, 2019
- Edmund S. Sabanegh, Jr. (20 October 2010). Male Infertility: Problems and Solutions. Springer Science & Business Media. pp. 83–. ISBN 978-1-60761-193-6
- O'Donnell L, Robertson KM, Jones ME, Simpson ER (2001). "Estrogen and spermatogenesis". Endocr. Rev. 22 (3): 289–318. doi:10.1210/edrv.22.3.0431. PMID 11399746

2

Approaches and Concepts of Developmental Biology

There are various approaches towards developmental biology such as anatomical approaches, experimental approaches and genetic approaches. This chapter closely examines these approaches towards developmental biology as well as key concepts related to it such as cell division, gene cloning and genetic mosaics.

Anatomical Approaches to Developmental Biology

Most of the questions in developmental biology have been bequeathed to it through its embryological heritage. There are numerous strands of embryology, each predominating during a different era. Sometimes they are very distinct traditions, and sometimes they blend. We can identify three major ways of studying embryology:

- Anatomical approaches
- Experimental approaches
- Genetic approaches

While it is true that anatomical approaches gave rise to experimental approaches, and that genetic approaches built on the foundations of the earlier two approaches, all three traditions persist to this day and continue to play a major role in developmental biology. In recent years, each of these traditions has become joined with molecular genetics to produce a vigorous and multifaceted science of developmental biology.

But the basis of all research in developmental biology is the changing anatomy of the organism. What parts of the embryo form the heart? How do the cells that form the retina position themselves the proper distance from the cells that form the lens? How do the tissues that form the bird wing relate to the tissues that form the fish fin or the human hand.

There are several strands that weave together to form the anatomical approaches to development. The first strand is comparative embryology, the study of how anatomy changes during the development of different organisms. For instance, a comparative embryologist may study which tissues form the nervous system in the fly or in the frog. The second strand, based on the first, is evolutionary embryology, the study of how changes in development may cause evolutionary changes and of how an organism's ancestry may constrain the types of changes that are possible. The third anatomical approach to developmental biology is teratology, the study of birth defects. These anatomical abnormalities may be caused by mutant genes or by substances in the environment that interfere with development. The study of abnormalities is often used to discover how normal development occurs. The fourth anatomical approach is mathematical modeling, which seeks to describe developmental phenomena in terms of equations. Certain patterns of growth and differentiation can be explained by interactions whose results are mathematically predictable. The revolution in graphics technology has enabled scientists to model certain types of development on the computer and to identify mathematical principles upon which those developmental processes are based.

Developmental Genetics

The function of genes is to pass on the information necessary to build proteins - and bodies - from one generation to the next. A newly fertilised egg cell has a collection of genes that contains all information needed to transform it from a single cell into an embryo and then an adult. The process that changes a single cell into a new person (or a new frog, or a new oak tree) is called development.

During the course of development, complex structures develop from simple ones. A single cell transforms itself into an adult organism. How something does complicated come from something simple? And how do genes control this process.

Creating an organism from a single cell involves three important processes:

- Cell division: Cells divide to produce more cells.
- Cell differentiation: Cells change into different types of cell to do specific jobs in the body, from nerve cells to muscle cells.
- Morphogenesis: Groups of cells move and change their shape to produce the structure of the organism.

Genes play a vital role in controlling all of these processes.

Genes contain the information a cell needs to make proteins - a bit like a recipe for a living thing. Different genes contain the information needed to make different proteins, and different proteins do different jobs in the cell. The proteins a cell makes decide what kind of cell it becomes, and there are some 350 different types of cell in an adult human being.

Cells change into different types of cell because of changes in the way their genes work. Some genes are activated (switched on), and some genes are inactivated (switched off). As a result, the cell produces a specific set of proteins. So, for example, a nerve cell produces only the proteins needed to make a nerve cell, and a muscle cell produces only the proteins needed to make a muscle cell.

But how do cells switch their genes on and off? And, more importantly, how do they 'know' which genes to switch on and which genes to switch off? The answer lies in special control genes that produce proteins that control the activity of other genes. So, for example, homeotic or homeobox genes control whole sets of other genes to set out the basic body plan of the embryo, separating the front from the back, and producing the right body structure in the right place.

One way in which genes can influence the activity of other genes is through the production of proteins called transcription factors, which stick to special control sites in the DNA at the start of a gene to switch them on and off.

Principles of Developmental Genetics

1. Differential gene expression from genetically identical nuclei creates different cell types. Differential gene expression can occur at the levels of gene transcription, nuclear RNA processing, mRNA translation, and protein modification.
2. Genes are usually repressed. Activation of a gene often means inhibiting its repressor. This leads to thinking in double and triple negatives: Activation is often the inhibition of the inhibitor; repression is the inhibition of the inhibitor of the inhibitor.
3. Eukaryotic genes contain promoter sequences to which RNA polymerase can bind to initiate transcription. The eukaryotic RNA polymerases are bound by a series of proteins called basal transcription factors.
4. Eukaryotic genes expressed in specific cell types contain enhancer sequences that regulate their transcription in time and space.
5. Specific transcription factors can recognize specific sequences of DNA in the promoter and enhancer regions. They activate or repress transcription from the genes to which they have bound.
6. Enhancers work in a combinatorial fashion. The binding of several transcription factors can act to promote or inhibit transcription from a certain promoter. In some cases transcription is activated only if both factor A and factor B are present, while in other cases, transcription is activated if either factor A or factor B is present.
7. A gene encoding a transcription factor can keep itself activated if the transcription factor it encodes also activates its own promoter. Thus, a transcription factor gene can have one set of enhancer sites to initiate its activation and a second set of enhancer sites (that bind the encoded transcription factor) to maintain its activation.
8. Often, the same transcription factors that are used during the differentiation of a particular cell type are also used to activate the genes for that cell type's specific products. For instance, Pax6 is needed both for the differentiation of the lens and for the transcription of the lens crystallin genes, and Mitf is needed for pigment cell differentiation and for the transcription of the genes whose products catalyze the synthesis of melanin.
9. Enhancers can act as silencers to suppress the transcription of a gene in inappropriate cell types.
10. Locus control regions may function by making relatively large portions of a chromosome accessible to transcription factors.
11. Transcription factors act in different ways to regulate RNA synthesis. Some transcription factors stabilize RNA polymerase binding to the DNA, some disrupt nucleosomes, and some increase the efficiency of transcription.

12. Transcription correlates with a lack of methylation on the promoter and enhancer regions of genes. Methylation differences can account for examples of genomic imprinting, wherein a gene transmitted through the sperm is expressed differently from the same gene transmitted through the egg.

13. Dosage compensation enables the X chromosome-derived products of males (which have one X chromosome per cell in fruit flies and mammals) to equal the X chromosome-derived products of females (which have two X chromosomes per cell). This compensation is accomplished at the level of transcription, either by accelerating transcription from the lone X chromosome in males (Drosophila) or by inactivating a large portion of one of the two X chromosomes in females (mammals).

14. X chromosome inactivation in placental mammals is generally random and involves the activation of the Xist gene on the chromosome that will be inactivated.

15. Differential RNA selection can allow certain transcripts to enter the cytoplasm while preventing other transcripts from leaving the nucleus.

16. Differential RNA splicing can create a family of related proteins by causing different regions of the nRNA to be read as exons and introns. What is an exon in one set of circumstances may be an intron in another.

17. Some messages are translated only at certain times. The oocyte, in particular, uses translational regulation to set aside certain messages that it transcribes during egg development but uses only after the egg is fertilized. This activation is often accomplished either by the removal of inhibitory proteins or by the polyadenylation of the message.

18. Many messenger RNAs are localized to particular regions of the oocyte or other cells. This localization appears to be regulated by the 3´ untranslated region of the mRNA.

Principles of Experimental Embryology

1. There are norms of reaction that describe an embryo's inherited ability to develop a range of phenotypes. The environment can play a role in selecting which phenotype is expressed. (Examples include temperature-dependent sex determination and seasonal phenotypic changes in caterpillars and butterflies.)

2. Developing organisms are adapted to the ecological niches in which they develop. (Examples include the ability of frog eggs exposed to sunlight to repair DNA damage.)

3. Before cells overtly differentiate into the many cell types of the body, they undergo a "covert" commitment to a certain fate. This commitment is first labile (the specification step) but later becomes irreversible (the determination step).

4. In autonomous specification, removal of a blastomere from an embryo causes the absence in the embryo of those tissues formed by that blastomere. This mechanism of specification

produces a mosaic pattern of development. (Examples include early snail and tunicate embryos.)

5. In autonomous specification, morphogenetic determinants are apportioned to different blastomeres during cell cleavage. (An example is the yellow crescent cytoplasm that is found in the muscle-forming cells of tunicate embryos.)
6. In conditional specification, the removal of a blastomere from the embryo can be compensated for by the other cells' changing their fates. Each cell can potentially give rise to more cell types than it normally does. This produces a regulative pattern of development wherein cell fates are determined relatively late. (Examples include frog and mammalian embryos.)
7. In conditional specification, the fate of a cell often depends upon its neighbors ("whom it meets").
8. In conditional specification, groups of cells can have their fates determined according to a concentration gradient of morphogen. The cells specified by such a morphogen can constitute a field.
9. In syncytial specification, the fates of cells can be determined by gradients of morphogens within the egg cytoplasm.
10. Different cell types can sort themselves into regions by means of cell surface molecules such as cadherins. These molecules can be critical in patterning cells into tissues and organs.

Cell Division

Cell division is the process cells go through to divide. There are several types of cell division, depending upon what type of organism is dividing. Organisms have evolved over time to have different and more complex forms of cell division. Most prokaryotes, or bacteria, use binary fission to divide the cell. Eukaryotes of all sizes use mitosis to divide. Sexually-reproducing eukaryotes use a special form of cell division called meiosis to reduce the genetic content in the cell. This is necessary in sexual reproduction because each parent must give only half of the required genetic material, otherwise the offspring would have too much DNA, which can be a problem.

Types of Cell Division

Prokaryotic Cell Division

Prokaryotes replicate through a type of cell division known as binary fission. Prokaryotes are simple organism, with only one membrane and no division internally. Thus, when a prokaryote divides, it simply replicates the DNA and splits in half. The process is a little more complicated than this, as DNA must first be unwound by special proteins. Although the DNA in prokaryotes usually exists in a ring, it can get quite tangled when it is being used by the cell. To copy the DNA efficiently, it must be stretched out. This also allows the two new rings of DNA created to be separated after they are produced. The two strands of DNA separate into two different sides of the prokaryote cell. The cell then gets longer, and divides in the middle. The process can be seen in the figure.

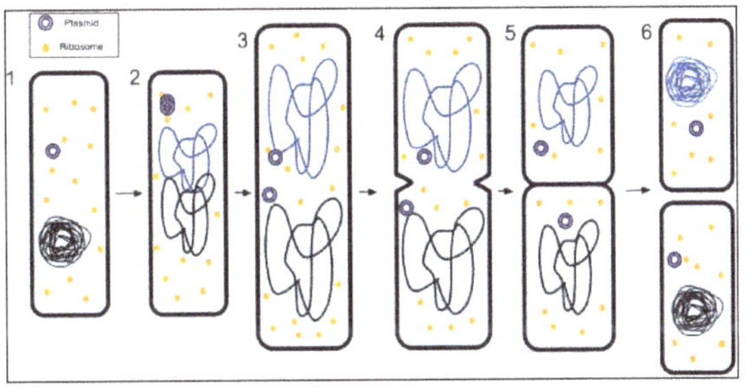

The DNA is the tangled line. The other components are labeled. Plasmids are small rings of DNA that also get copied during binary fission and can be picked up in the environment, from dead cells that break apart. These plasmids can then be further replicated. If a plasmid is beneficial, it will increase in a population. This is in part how antibiotic resistance in bacteria happens. The ribosomes are small protein structures that help produce proteins. They are also replicated so each cell can have enough to function.

Eukaryotic Cell Division: Mitosis

Eukaryotic organisms have membrane bound organelles and DNA that exists on chromosomes, which makes cell division harder. Eukaryotes must replicate their DNA, organelles, and cell mechanisms before dividing. Many of the organelles divide using a process that is essentially binary fission, leading scientist to believe that eukaryotes were formed by prokaryotes living inside of other prokaryotes.

After the DNA and organelles are replicated during interphase of the cell cycle, the eukaryote can begin the process of mitosis. The process begins during prophase, when the chromosomes condense. If mitosis proceeded without the chromosomes condensing, the DNA would become tangled and break. Eukaryotic DNA is associated with many proteins which can fold it into complex structures. As mitosis proceeds to metaphase the chromosomes are lined up in the middle of the cell. Each half of a chromosome, known as sister chromatids because they are replicated copies of each other, gets separated into each half of the cell as mitosis proceeds. At the end of mitosis, another process called cytokinesis divides the cell into two new daughter cells.

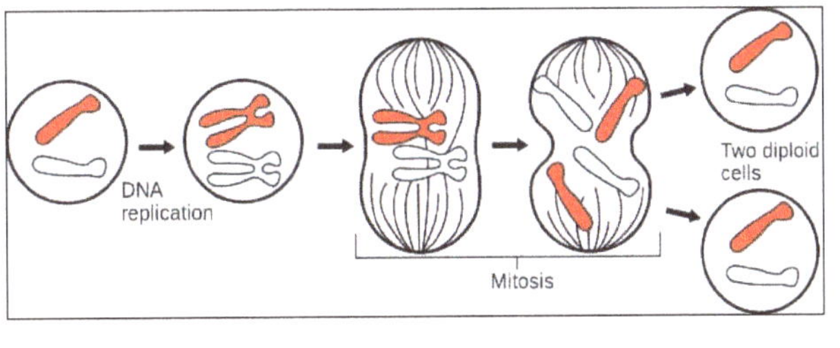

All eukaryotic organisms use mitosis to divide their cells. However, only single-celled organisms use mitosis as a form of reproduction. Most multicellular organisms are sexually reproducing and combine their DNA with that of another organism to reproduce. In these cases, organisms need a different method of cell division. Mitosis yields identical cells, but meiosis produces cells with half the genetic information of a regular cell, allowing two cells from different organisms of the same species to combine.

Eukaryotic Cell Division: Meiosis

In sexually reproducing animals, it is usually necessary to reduce the genetic information before fertilization. Some plants can exist with too many copies of the genetic code, but in most organisms it is highly detrimental to have too many copies. Humans with even one extra copy of one chromosome can experience detrimental changes to their body. To counteract this, sexually reproducing organisms undergo a type of cell division known as meiosis. As before mitosis, the DNA and organelles are replicated. The process of meiosis contains two different cell divisions, which happen back-to-back. The first meiosis, meiosis I, separates homologous chromosomes. The homologous chromosomes present in a cell represent the two alleles of each gene an organism has. These alleles are recombined and separated, so the resulting daughter cells have only one allele for each gene, and no homologous pairs of chromosomes. The second division, meiosis II, separated the two copies of DNA, much like in mitosis. The end result of meiosis in one cell is 4 cells, each with only one copy of the genome, which is half the normal number.

Organisms typically package these cells into gametes, which can travel into the environment to find other gametes. When two gametes of the right type meet, one will fertilize the other and produce a zygote. The zygote is a single cell that will undergo mitosis to produce the millions of cells necessary for a large organism. Thus, most eukaryotes use both mitosis and meiosis, but at different stages of their lifecycle.

Cell Division Stages

Depending upon which type of cell division an organism uses, the stages can be slightly different.

Mitosis Stages

Mitosis starts with prophase in which the chromosome is condensed. The cell proceeds to metaphase where the chromosomes are aligned on the metaphase plate. Then the chromosomes are separated in anaphase and the cell's cytoplasm is pinched apart during telophase. Cytokinesis is the final process that breaks the cell membrane and divides the cell into two.

Meiosis Stages

The stages of meiosis are similar to mitosis, but the chromosomes act differently. Meiosis has two phases, which include two separate cell divisions without the DNA replicating between them. Meiosis I and meiosis II have the same 4 stages as mitosis: prophase, metaphase, anaphase, and telophase. Cytokinesis concludes both rounds of meiosis.

In prophase I, the chromosomes are condensed. In metaphase I, the chromosomes line up across from their homologous pairs. When they are separated in anaphase I and telophase I, there is only one form of each gene in each cell, known as a reduction division. Meiosis II proceeds in the same manner as mitosis, which sister chromatids dividing on the metaphase plate. By telophase II, there are 4 cells, each with half of the alleles as the parent cell and only a single copy of the genome. The cells can now become gametes and fuse together to create new organisms.

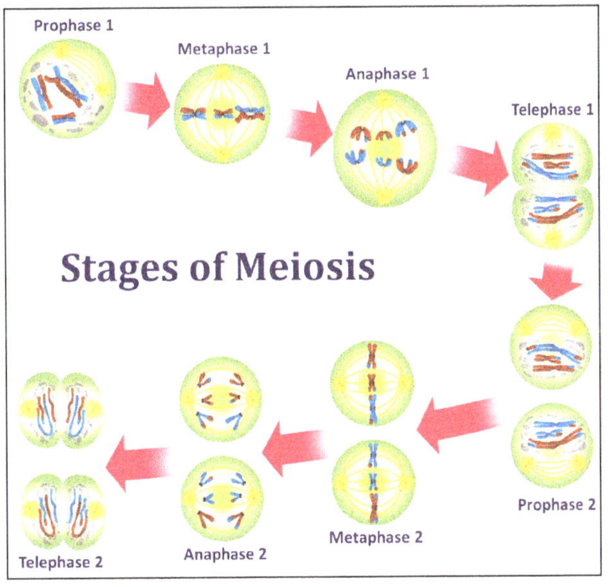

Gene Cloning

The production of exact copies (clones) of a particular gene or DNA sequence using genetic engineering techniques. The DNA containing the target gene(s) is split into fragments using restriction enzymes. These fragments are then inserted into cloning vectors, such as bacterial plasmids or bacteriophages, which transfer the recombinant DNA to suitable host cells, such as the bacterium E. coli. Alternatively, complementary DNA is inserted into the vectors, or 'naked' DNA fragments can be taken up directly by a host bacterium from its medium (this is less efficient than vector transfer).

Inside the host cell the recombinant DNA undergoes replication; thus, a bacterial host will give rise to a colony of cells containing the cloned target gene. Various screening methods may be used to identify such colonies, enabling them to be selected and cultured. Gene cloning facilitates DNA sequencing; it also enables large quantities of a desired protein product to be produced. Human insulin, for example, is now produced by bacteria containing the cloned insulin gene.

Gene Duplication and Divergence

Gene duplication, followed by divergence is a source of new genes in genomes. Divergence literally means to go in different directions. In this context, we mean that the sequences of the gene copies are becoming different from each other because of the accumulation of mutations. As you go through the worksheet, you will understand how this process works.

Genes Duplication

Genes can get duplicated when there is unequal crossing over of chromosomes. (There are also other known mechanisms, but we did not discuss these.) As you know, during meiosis, homologous chromosomes often exchange equivalent regions. This is called recombination or crossing over.

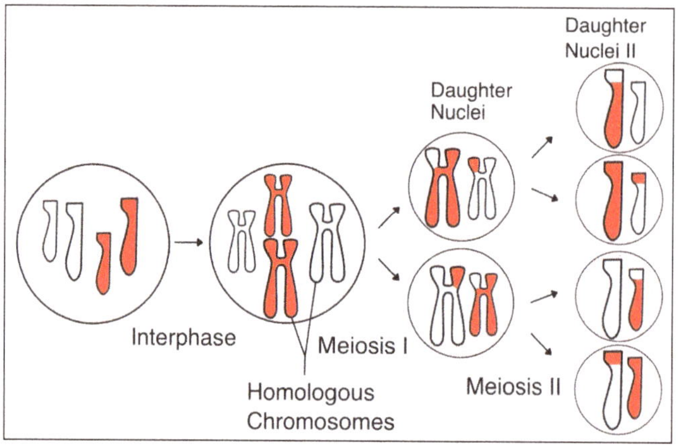

Notice that in the figure above, the chromosomes are lined up so that they exchange equivalent regions. This means that neither chromosome involved in the crossover gains or loses any information, they simply swap the equivalent sequence from one to the other. Note also, that after meiosis is complete, there are four gametes, each of which has a single copy of each chromosome.

In figure shows chromosomes lined up normally, where any crossing over event will result in the exact exchange of equivalent regions between the two chromosomes. For simplicity, only one pair of chromosomes is shown here, rather than the entire tetrad. Notice that the same repeated sequence, represented by ATAT, is present on either side of the gene.

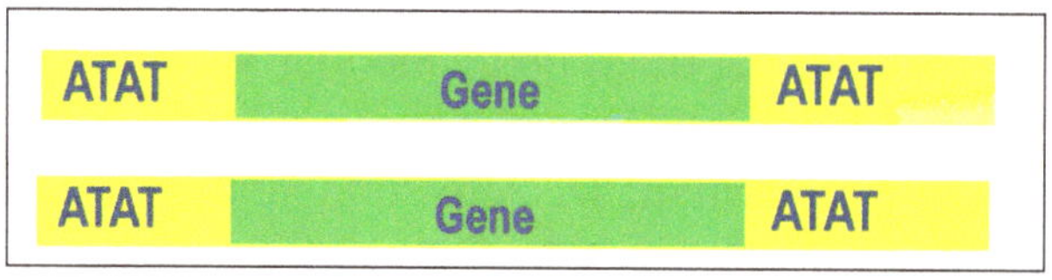

In figure shows the same chromosomes, but they are not lined up correctly. However, because they have the same repeated sequences on either side of the gene of interest; it is still possible for them to line up at a different repeat that is offset.

Draw out what happens when these two chromosomes swap sections by crossing over at the offset ATAT repeat shown in the figure. Each of the resulting chromosomes would go into a gamete (an egg or a sperm). Because only two chromosomes are shown here, you get only 2 gametes, instead of the four that would result from a tetrad in meiosis

Gene duplication can happen multiple times: Leading to multiple copies of the same gene on a single chromosome. This happens by a repetition of the process above. You may draw out for yourself how further duplication could result in more copies of the gene on the chromosome.

Divergence of the Copies

Once you have more than one copy of a gene on a single chromosome, it is possible for the copies to gradually become different, as the gene copy (or copies) undergoes mutation. Mutations in these copies are not detrimental to the organism, as long as there is at least one "good" original copy. Because of this, the mutant copies can be passed on to the offspring. Remember, mutations are not always harmful, so sometimes, the mutation will result simply in a slight variation in the properties of the encoded protein. This can turn out to be beneficial to the organism if the variant protein confers an advantage of some kind.

On the figure below with two gene copies, mark one copy a and one copy a'. To start with, both these are identical copies. But as one copy gets mutated, it may become sufficiently different that it encodes a slightly different protein. You would then recognize it as a new, though related, gene and call it b.

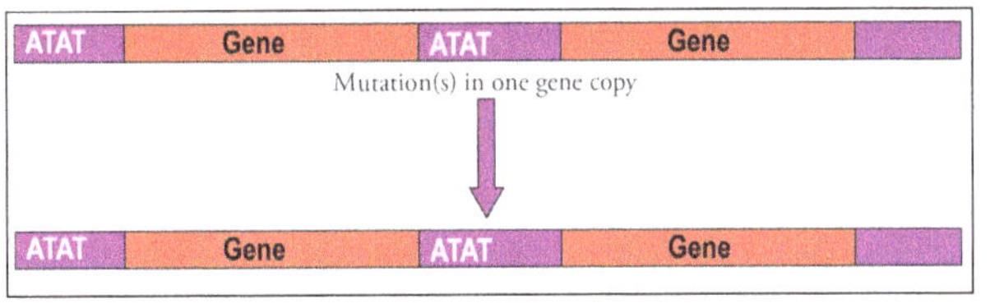

You should now have one original copy of the gene, a and one variant, b, that give rise to two slightly different proteins. The accumulation of differences (caused by mutation) in the genes is what we call "divergence".

The process above is how the original single globin gene is thought to have given rise to two genes that became the α and β globin genes. Subsequent duplication and divergence in these two genes gave rise to the variant forms of α and β that are now seen in humans. In the figure below, you can see that the original single globin gene that was present in early vertebrates was duplicated and gave rise, initially, to the α and β forms. The figure shows the variants that arose in the β gene, but not the α gene. One of the variant forms of the β is the fetal γ form that we discussed in class, that gives rise to the a_2g_2 hemoglobin that binds oxygen with greater affinity than the adult a_2b_2 form.

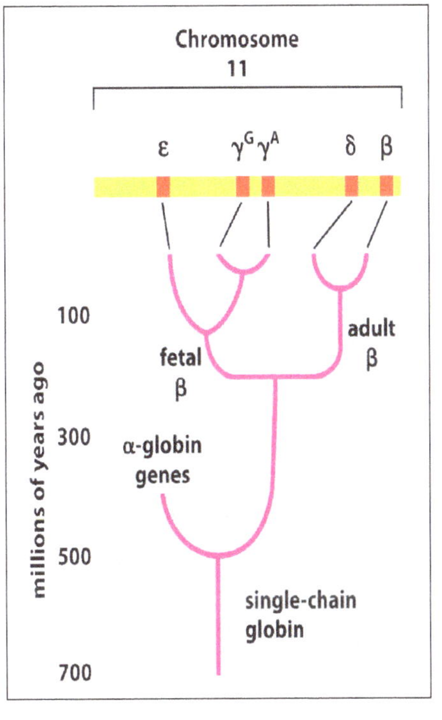

The collection of different globin genes that arose from the common ancestral globin gene are called the globin gene family. Like human families, a gene family has related, but recognizably different members. Also like human families, gene families may have dysfunctional members. Such dysfunctional genes are gene copies that have mutations that render them non-functional. Because they look like other members of their gene family, but do not give rise to functional proteins, such non-functional gene family members are called pseudogenes.

There are thousands of gene families within the human genome. Like the globin gene family, each of these families is made up of related but slightly different members that arose from an ancestral form. One example is the histone gene family that gives rise to the various different histone proteins that you are familiar with.

Gene Families

Comparison of the genome sequences of different organisms shows us how genomes have changed through time. As we noted, early vertebrates had a single globin gene. More recent descendants of these organisms have more than one globin gene, but these are clearly derived from the original gene, as we can see from the sequence. The figure above shows at what stage in the history of life these different variant forms appeared. Although the globin genes have undergone change over millions of years, there are also recent examples of more rapid changes in genomes that confirm that this mechanism is responsible for the emergence of gene families. For example, scientists have observed at least three independent gene duplication events in the acetylcholine esterase genes of certain mosquitoes over the past 50 years. These duplications have been followed by selection for variant forms of these genes that make the mosquitoes resistant to insecticides. Thus, in a relatively short period of time, mosquitoes have acquired extra copies of the ace-1 gene, some of which have mutations that enable the mosquitoes to survive better in the presence of certain insecticides.

Genetic Mosaics

A genetic mosaic is an individual in which different cells have different genotypes. In Caenorhabditis elegans, genetic mosaics have been generated, identified, and analyzed for the purpose of defining where in the animal a gene of interest must be expressed to cause a particular phenotypic effect. Most C. elegans mosaics that have been studied were homozygous mutant for a recessive mutation in all cells and carried additionally, in some cells, one or more copies of the wild-type allele of the same gene on an extra chromosomal element. Before mosaic analysis is undertaken, a mutant phenotype corresponding to the absence of the wild-type gene from all cells is analyzed and described. Some mutant phenotypes are described in whole-animal terms, such as uncoordinated movement, longer-than-normal lifespan, abnormal body shape, or in viability. Such traits can be analyzed in mosaic animals. The overall phenotypes are recorded for mosaic animals in which it is known which cells carry the wild-type gene and which do not, with the goal of establishing which cells must carry the wild-type gene to prevent the appearance of the mutant phenotype. The responsible cell or group of cells is referred to as the anatomical focus of the action of the gene with respect to the phenotype under study.

In a second class of mutants, it is known at the outset that particular cells—specific nerve or muscle cells, for example—exhibit an abnormality, which is generally detected microscopically. Mosaic animals can then be used to ask whether or not the cellular focus of action with respect to this phenotype is in the very cell or cells that exhibit the abnormality; this is equivalent to asking whether or not the gene acts cell autonomously with respect to the cellular phenotype. When a gene acts cell autonomously, the phenotype of a cell is determined solely by the genotype of that cell and not by the genotypes of neighboring or other cells. A gene is said to act cell non-autonomously when a genotypically homozygous mutant cell exhibits a wild-type phenotype or when a cell carrying a wild-type allele exhibits a mutant phenotype. Cell non-autonomy is caused by intercellular interactions, and mosaic analysis can help elucidate the nature of the interaction. For example, a gene that produces a ligand that must be recognized by a receptor on a second cell could act cell non-autonomously with respect to a phenotypic effect in the receiving cell.

Generation of Worm Genetic Mosaics

The generation, identification, and analysis of genetic mosaics in C. elegans have recently been reviewed. Also, a few of the technical details concerning mosaics are mentioned elsewhere in Worm Book. Thus, these issues shall only be briefly mentioned here. Worm genetic mosaics have usually been generated from zygotes that were homozygous mutant for the gene under study and carried the wild-type allele of the same gene on an extra chromosomal element—either an extra chromosomal array or a free duplication of part of a chromosome—that was subject to spontaneous mitotic loss. If the extra chromosomal element is present in all cells, the animal exhibits a wild-type phenotype. But if the extra chromosomal element fails to be transmitted to a daughter cell during the development of the animal, all of the descendants of that cell, a clone, will lack the wild-type allele and be homozygous mutant. The extra chromosomal element should carry a second gene that behaves cell autonomously and can be scored to track which cells carry the extra chromosomal element and which do not. The well-described and nearly invariant cell lineage is a great aid in determining where in the lineage the extra chromosomal element was lost. This allows one to determine precisely, without scoring every cell, the genotypes of essentially every cell in the mosaic animal. C. elegans mosaics have also been generated by other methods, and chimeric partial embryos have been formed in vitro from EMS and P_2 blastomeres for genetic studies of Wnt signaling during early development.

Examples of Mosaic Analysis in Worms

Over 75 genes have been analyzed in C. elegans genetic mosaics. Several examples of how mosaic analysis has contributed important insights into worm development were recently reviewed. Included were genes implicated in cell-to-cell signaling, with the functions of some genes assigned to the signaling cell and the functions of others assigned to the signal-receiving cell. Other examples in which anatomical foci were determined included genes affecting whole-animal phenotypes such as uncoordination, body size, and lifespan.

Mutations that confer uncoordinated movement often affect the body-wall muscles or the nervous system. The nature of the cell lineage makes it possible quickly to distinguish these candidate cell types: all of the body neurons, except for two, descend from AB, a daughter of the zygote; and all of the body-wall muscles, with the exception of one, descend from P_1, the other daughter of the zygote. A thorough analysis of mosaics is, or course, required for rigorous determination of the actual focus of activity of the gene, which could be in a cell type other than muscle or neuron. In a recent analysis, the anatomical focus of unc-122 has been traced to muscle. The unc-71 gene, which regulates both motor-axon guidance and sex-myoblast migration, is thought to act cell nonautonomously for both functions. The gene is required in both neurons and epidermis for proper motor-axon guidance.

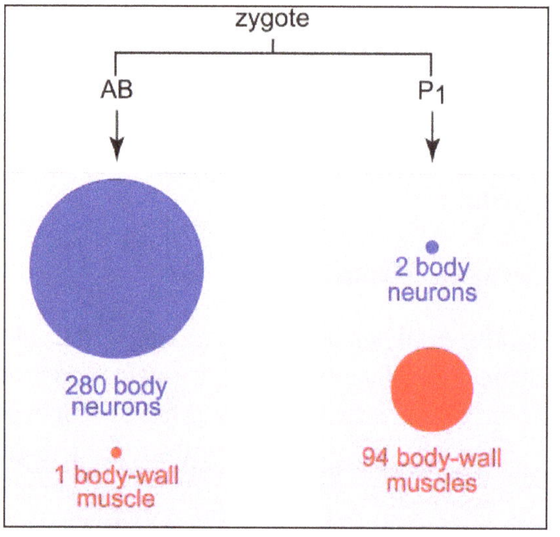

AB-P_1 mosaics with respect to body neurons and body-wall muscle the separation of the precursors for most of the body neurons from those of most of the body-wall muscle occurs at the first division of the zygote.

A problem for developmental studies is the use of signal-transduction molecules in several distinct processes. In some cases, later requirements for such molecules can be masked by a critical requirement for the molecules earlier in development. Mosaics can in some cases uncover the foci of activity of the gene in these various processes. A good example is mpk-1, which encodes an ERK2-type MAP-kinase known to affect vulval development. For proper vulval development, mpk-1 is required in the vulval precursor cells. mpk-1 is also known to affect the exit of germ nuclei from the pachytene stage of meiosis, and in this case, the focus is the germ line itself. The focus of a vital function for mpk-1 has now been shown to be hyp7, the large syncytium that composes most of the skin of the animal.

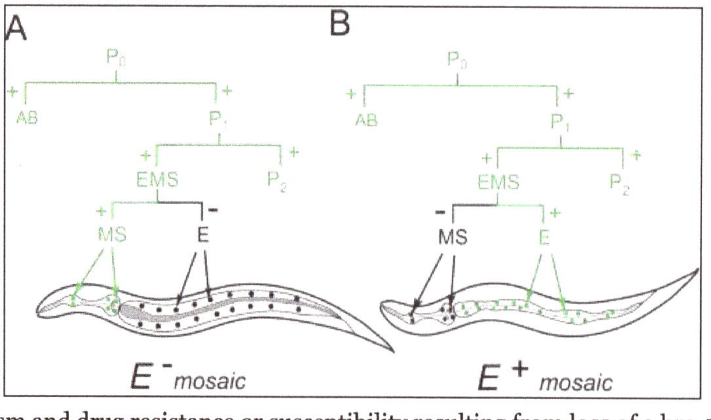

Mosaicism and drug resistance or susceptibility resulting from loss of a bre-5+ extra chromosomal array during division of the blastomere EMS in an otherwise bre-5 mutant background.

Illustrating the power of mosaics for non-developmental studies is bre-5, mutation of which confers resistance to a toxin from Bacillus thuringiensis. Animals expressing the dominant wild-type gene (conferring sensitivity to toxin) in the intestine were sensitive to the toxin, whereas 67 mosaics that were resistant to the toxin were always homozygous mutant in all of the cells of the intestine. The bacterial toxin appears to be harmful to C. elegans only if it can be ingested, which may require binding to a carbohydrate expressed by the intestinal cells. bre-5, which encodes a β-1,3-galactosyltransferase, may be required for synthesis of the carbohydrate.

In figure, inheritance of the array during development is inferred from expression of a green marker also present on the array. During development of the embryo in A, the array was inherited by MS but not by E, its sister. MS, the progenitor of several mesodermal cell types (a few in the pharynx are indicated), is not a progenitor of intestinal cells. They, instead, all derive from E, which only gives rise to intestinal cells. Mosaics, such as the example in A, that have not inherited the array in the intestine (black nuclei), are resistant to the Bacillus thuringiensis toxin, indicating that bre-5 is required in the intestine for susceptibility . In contrast, the intestinal cells in mosaic B should express wild-type bre-5, because they have inherited the array (green nuclei). The presence of wild-type bre-5 in the gut makes the animal competent to ingest the toxin, resulting in intestinal pathology.

Continued Importance of Mosaic Analysis

In expression studies, some have no doubt encountered the situation in which a reporter construct rescued the mutant phenotype, but expression of the reporter gene could not be reliably detected. In some cases, antibodies against the reporter can provide increased sensitivity, but the concern remains that expression of a reporter construct gives spurious results, as mentioned above. Mosaic analysis, on the other hand, provides a functional test of the gene that does not depend on the ability to detect a reporter. Although there are potential limitations, as discussed above, mosaic analysis in many cases can provide an unbiased analysis of the activity of a gene, and the analysis can provide insight into the function of genes that are not easily amenable to immunocytochemical and other studies, because they are expressed at low levels in only a few cells.

A maternal contribution of a gene can be assessed when germ-line clones are generated following the loss of an extra chromosomal element in P_1, P_2, P_3, or P_4, the precursors of the germ cells .A maternal contribution is evident when the progeny of a germ-line mosaic have a more severe phenotype than progeny that segregate from a mother who is heterozygous for the gene A full-fledged mosaic analysis is not necessary for an analysis of germ-line clones. They can be seen in the course of maintaining lines rescued by an extra chromosomal element. If many individuals are put one to a plate, a few of the plates may have received a germ-line mosaic. For a gene with a completely penetrant phenotype, all of the progeny will have a mutant phenotype. A maternal contribution can also be assessed with RNA interference (RNAi), although the potential for cross-interference and for incomplete depletion of gene product should be kept in mind.

Another method for asking about the anatomical focus of the activity of a gene is to introduce transgenic copies of the wild-type gene controlled by cell- or tissue-specific promoters into otherwise homozygous mutant animals. Rescue of a mutant defect by a particular transgene indicates that expression in the indicated cell type can provide the needed function. There are, however, some concerns when using this approach. One is the fidelity of expression of such transgenes, especially from arrays that contain multiple copies of the engineered gene and other heterologous DNA. Also, a bit of gene cloning is involved, because a new construct must be made for each promoter under study. The formal possibility also remains that rescue by expression (or overexpression) from a transgene does not prove that the wild-type gene normally acts in the same cell type or tissue.

References

- Developmental-genetics: le.ac.uk, Retrieved 9 May, 2019
- Cell-division: biologydictionary.net, Retrieved 11 June, 2019
- Gene-cloning, science: encyclopedia.com, Retrieved 13 July, 2019

3

Development of Organs

The phase of embryonic development that starts at the end of gastrulation and lasts till birth is known as organogenesis. This chapter has been carefully written to provide an easy understanding of the different processes related to development of organs such as organogenesis and cellular differentiation.

Organogenesis

Organogenesis is the process by which the three germ tissue layers of the embryo, which are the ectoderm, endoderm, and mesoderm, develop into the internal organs of the organism. Organs form from the germ layers through the differentiation: the process by which a less-specialized cell becomes a more-specialized cell type. This must occur many times as a zygote becomes a fully-developed organism. During differentiation, the embryonic stem cells express specific sets of genes which will determine their ultimate cell type. For example, some cells in the ectoderm will express the genes specific to skin cells. As a result, these cells will differentiate into epidermal cells. Therefore, the process of differentiation is regulated by cellular signaling cascades.

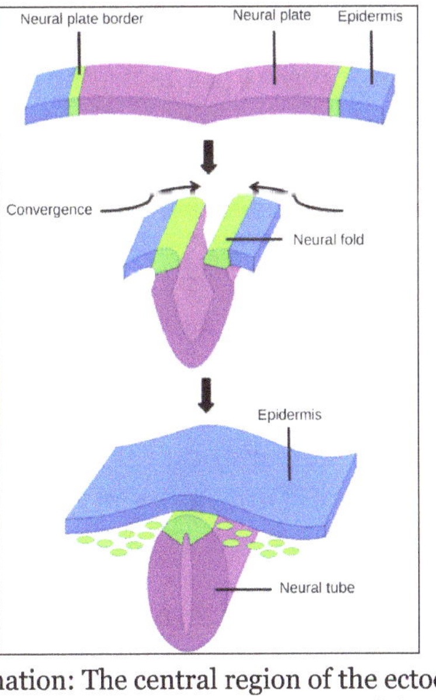

Neural tube formation: The central region of the ectoderm forms the neural tube, which gives rise to the brain and the spinal cord.

In vertebrates, one of the primary steps during organogenesis is the formation of the neural system. The ectoderm forms epithelial cells and tissues, as well as neuronal tissues. During the formation of the neural system, special signaling molecules called growth factors signal some cells at the edge of the ectoderm to become epidermis cells. The remaining cells in the center form the neural plate. If the signaling by growth factors were disrupted, then the entire ectoderm would differentiate into neural tissue. The neural plate undergoes a series of cell movements where it rolls up and forms a tube called the neural tube. In further development, the neural tube will give rise to the brain and the spinal cord.

The mesoderm that lies on either side of the vertebrate neural tube will develop into the various connective tissues of the animal body. A spatial pattern of gene expression reorganizes the mesoderm into groups of cells called somites, with spaces between them. The somites will further develop into the ribs, lungs, and segmental (spine) muscle. The mesoderm also forms a structure called the notochord, which is rod-shaped and forms the central axis of the animal body.

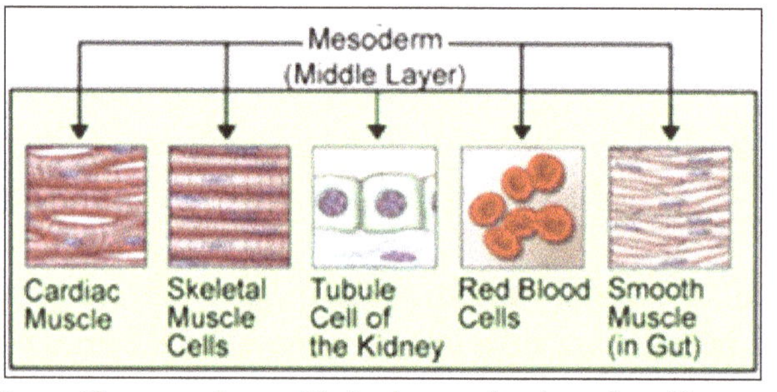

Mesoderm: The mesoderm aids in the production of cardiac muscles, skeletal muscle, smooth muscle, tissues within the kidneys, and red blood cells.

The endoderm consists, at first, of flattened cells, which subsequently become columnar. It forms the epithelial lining of the whole of the digestive tube (except part of the mouth and pharynx) and the terminal part of the rectum (which is lined by involutions of the ectoderm). It also forms the lining cells of all the glands which open into the digestive tube, including those of the liver and pancreas; the epithelium of the auditory tube and tympanic cavity; the trachea, bronchi, and air cells of the lungs; the urinary bladder and part of the urethra; and the follicle lining of the thyroid gland and thymus. Additionally, the endoderm forms internal organs including the stomach, the colon, the liver, the pancreas, the urinary bladder, the epithelial parts of trachea, the lungs, the pharynx, the thyroid, the parathyroid, and the intestines.

Organs are usually defined as containing two or more tissues, each tissue containing differing cell types and cell functions, coordinated in a higher level of organization and function than the independent tissues. Second to the organism's overall body organization, organs are the most complex level of organization of cells. Organogenesis is the organ-forming phase of embryonic development. It begins once the basic anteroposterior and dorsoventral organization of the embryo is established by gastrulation and neurulation. During organogenesis, cytodifferentiation takes place, and then the organ begins to function.

A fundamental question about organogenesis concerns the means by which the different parts of the organ are brought into complex alignment and integrated function. In the first half of the twentieth century, organ formation was described in detail by light microscopy, and the inductive interactions of different cell groups involved in organ formation were revealed by experimental analysis. In general, the different tissues of the organ were not found to form independently and then come together in perfect apposition. Rather, tissues that are nearby as a result of extensive movement during gastrulation and neurulation interact with each other and also with surrounding tissues. Combinations of signals establish positional identity and initiate the progressive delineation of organ-specific gene activations. Thus, it is not necessary that all participants in early organogenesis have position and cell-type specific information. Cell signaling operates throughout organogenesis. Recently, the local signals and responses have been identified in several kinds of organogenesis, the responses often being experimentally proven by using "marker" or "reporter" genes activated at various stages of the process and visualized by staining specific mRNAs by in situ hybridization. Extensive molecular descriptions and cellular and genetic analyses have defined key regulatory pathways that facilitate the development of many vertebrate systems, including the following:

- Neural tube: Regionalization of forebrain, midbrain, hindbrain, and spinal cord.
- Neural tube: Dorso-ventral organization of brain and spinal cord.
- Sensory systems: Optic vesicle and eye, otic vesicle and inner ear, and olfactory epithelium.
- Neural crest: Autonomic and sensory ganglia and glia and melanocytes.
- Neural crest: midfacial and branchial connective tissues and teeth.
- Paraxial mesoderm: Somites, skeletal muscle, vertebrae, and ribs.
- Intermediate mesoderm: Kidneys, gonads, reproductive ducts, and sex.
- Cardiovascular system: Heart, angiogenesis, and hematopoiesis.
- Limb: Growth and specification of axes.
- Pharyngeal endoderm: Thyroid and thymus.
- Gut tube: Lungs, liver, pancreas, stomach, and intestines.

The familiar conserved signaling pathways are used over and over in many different contexts in organogenesis and other steps of development. For example, the Sonic Hedgehog (SHH) signaling pathway is involved in establishing asymmetry in the early gastrula, inducing floor plate and motor neurons, separating the single eye field into paired optic primordia, maintaining proliferation in migrating neural crest cells, establishing patterning of the medial and lateral nasal prominences and tooth induction, inducing sclerotome segregation and epaxial muscle formation in somites, development of the prostate gland, determining left-right asymmetry, establishing the anteroposterior (rostrocaudal) axis of the limbs, delineating the tracheo-esophageal diverticulum, and establishing sites of formation and branching patterns in lung and pancreatic epithelia. Comparable matches could be made for members of the fibroblast growth factor (FGF), TGFβ, BMP, and WNT signaling families. Although the signaling pathways involve the same or closely related signaling molecules, the responses made by cells are distinct because of the genes and gene products they express prior to and in response to the many different combinations of these signaling factors.

Vertebrate Limb: The Best Known Organogenesis Model

The limb is by far the most studied organ rudiment of vertebrates, supported by over 50 years of experimental embryology and intensive recent molecular genetic analysis. Cell-cell interactions, based on known signaling pathways, are well understood, as are the patterns of cell proliferation within the developing limb. Surprisingly, many homologous genes and signaling pathways are conserved between the vertebrate limb and the developing leg or wing of Drosophila. The researchers who followed the seemingly remote leads from the early Drosophila work on appendages have made rapid progress on vertebrate limb development in the past few years.

Limb development is: (1) the importance of precise temporal and spatial signaling for organizing a complex organ rudiment, and (2) the interaction of multiple signaling pathways to establish the organ's three-dimensional morphology.

The developing limb contains three axes: proximodistal, anteroposterior (rostrocaudal), and dorsoventral. Each axis has its own secreted signals and these are integrated in the limb bud, The bud originates as a locus of rapidly dividing cells in the somatic mesoderm and overlying epidermal ectoderm of the flank of the trunk. Outgrowth of the paired anterior and posterior limbs involves the maintenance of a high rate of cell division within the bud. Local trunk structures, such as the mesonephros, somites, and notochord, secrete FGF10 onto the flank tissues, and this signal initially keeps the bud proliferating. As outgrowth begins, the ectoderm overlying the bud locally thickens to form a ridgelike structure called the apical ectodermal ridge (AER), which then secretes several FGFs, notably FGF8, and takes over for the flank in maintaining cell proliferation in the adjacent bud mesoderm. The bud is then self-sufficient for this signaling and the flank stops serving as an FGF source. (The bud can be transplanted to a remote site, such as the yolk sac, at this stage and will develop autonomously.) The area of rapid division in the bud is called the progress zone (PZ). It lies just beneath the AER. As cells proliferate in it, some are displaced away from the AER and are no longer exposed to FGF. They slow their division rate and stop changing their specification (i.e., their capacity to develop as one part or another of the future limb). As a consequence, cells leave the PZ in a proximal to distal order of specification (i.e., the first to leave have the capacity to form the upper limb, the next to form the lower limb, and the last to form the hand or foot). When cells leave the PZ, they start to differentiate into cartilage, bone, and connective tissue of their specified limb level. Precursor cells of limb muscles migrate into the bud from the adjacent somitic myotomes, and nerves extend in from spinal ganglia and the spinal cord. Interactions between the AER and PZ are reciprocal (i.e., the AER maintains mitosis within the PZ by way of FGF, and the PZ maintains the thickened AER by way of a yet unknown signal).

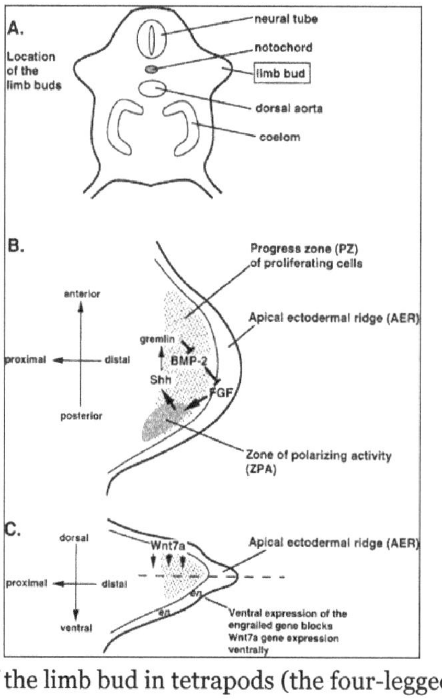

Development of the limb bud in tetrapods (the four-legged vertebrates).

In figure, (Panel A) Cross section of mouse embryo after ventral closure, at the level of the forelimbs. The limb buds emerge as small protrusions from the left and right flank. They consist of a surface layer of epidermal cells (ectoderm) and an internal mass of mesenchymal cells (mesoderm). The latter engage in rapid proliferation. (Panel B) Close-up view of a bud, cross section, dorsal, ventral. The apical ectodermal ridge (AER) secretes FGF onto the underlying mesenchyme. The zone of polarizing activity (ZPA) of the mesenchyme secretes SHH onto the ridge. A reciprocal activation circuit is completed in which the AER and ZPA keep each other active. In the mesenchyme close to the ridge is the progress zone (PZ), a population of rapidly dividing cells. Their division is kept going by FGF and SHH. Dividing cells keep changing their option for a future developmental path, in a progression, proximal (shoulder) to distal (hand). As they divide, some are displaced toward the flank, out of the PZ, too far away to receive FGF signals. Their division slows, and they adopt the developmental path of the step of the progression at which they were when they were last in the PZ. The SHH from the ZPA spreads in a gradient toward the anterior edge of the bud. The antero-posterior differences of the limb are signaled by this gradient, perhaps through a BMP2,4 signaling coupled to SHH signaling. (Panel C) The limb bud in cross section, at right angles to panel B, to show the dorsoventral plane. The dorsal epidermis secretes WNT7A onto the mesenchyme. That signals it to develop dorsally. The ventral epidermis does not secrete WNT7A, because the cells express the En gene encoding a transcription factor inhibiting the Wnt7a gene from expression.

The second axis of the limb, the anteroposterior axis, is also established through cell-cell interactions and secreted signals. The posterior portion of the limb bud contains a specialized region called the zone of polarizing activity (ZPA). This zone was originally recognized as a signaling center, because when it was transplanted to the anterior side of another limb bud, the bud develops a mirror-image duplicated limb. The ZPA secretes SHH, which diffuses across the anteroposterior dimension of the bud, establishing a gradient and setting off the formation of a second gradient of BMP2 and 4, two kinds of TGFβ signals. SHH is both sufficient and necessary to establish the anteroposterior pattern of the limb. Retinoic acid, acting via a nuclear receptor, might also have a role.

The third axis of the limb, the dorsoventral axis, is established through interactions between the nonridge dorsal ectoderm of the limb and the underlying bud mesoderm. That was initially shown by experiments in which the dorsal and ventral regions of the ectoderm were rotated with respect to the mesoderm. The dorsal ectoderm releases the WNT7A signal, which induces the expression of the Lmx-1 gene in the underlying mesoderm and which suppresses expression of the engrailed-1 gene, thereby restricting its expression to the ventral ectoderm. Mutants defective in WNT7A signaling develop double ventral limbs. Mutants defective in engrailed-1 expression develop double dorsal limbs with double sets of fingernails.

It remains to be learned how the signaling pathways of each axial dimension are coordinated with those of the other dimensions, and how the integration of these pathways leads to the formation of unique skeletal structures in precise locations within the limb. The limb exemplifies the advanced understanding of vertebrate organogenesis at a molecular genetic level (i.e., of the signal pathways and the genetic regulatory circuits involved in changing transcription and regulating cell proliferation). This kind of understanding is prerequisite to understanding the action of toxicants on embryogenesis. On the basis of new information, mechanisms of action of toxicants have been recently proposed, although these have yet to be tested. For example, thalidomide leads to a failure to form proximal parts of the limb (the upper and middle parts of the limb), but the hand or foot is usually formed. That developmental outcome is paradoxical, because other treatments, such as the removal or inactivation of the AER, lead to truncation of the limb in the reverse order; the upper and middle parts are present but the hand or foot is missing. Tabin recently proposed that thalidomide reduces cell proliferation in the PZ, by means yet unknown, and the few cells that remain there in prolonged contact with FGF8 secreted by the AER are specified as hand or foot cells, the last normally to emerge from the zone. This is an example of an incisive prediction about a long-known toxicant made on the basis of recent knowledge. Yet, the chemical basis for thalidomide's specific effect on cell proliferation in the PZ escapes even a proposal at this time, and so the hypothesis is incomplete. Stephens and Fillmore have suggested that thalidomide interferes with integrin gene expression in limb bud mesenchyme cells and, thereby, with their ability to stimulate angiogenesis at the level needed for continued rapid proliferation.

The capacity of the limb bud to develop normally after injury has been studied. Large numbers of cells can be removed at early stages, and as long as representatives remain of the AER, ZPA, and PZ, development will be normal. Immigrating muscle cells from any myotome will enter and adapt to the limb bud, and nerves from any spinal cord level will enter and make neuromuscular connections, although the CNS circuitry of that level may not be appropriate for normal limb movement. The robustness of limb development, like that of other parts of the body, is substantial, because each of the interactive cell groups is much larger than minimally necessary and is capable of proliferation. Robustness is not unlimited, however, and total removal of a key signaling or responding group is deleterious. Regenerating limbs, such as those of newts, have surmounted even that limit, but they are the exception among vertebrates.

Organs Produced by the Germ Layers

The endoderm is the inner most germ layer of the embryo which gives rise to gastrointestinal and respiratory organs by forming epithelial linings and organs such as the liver, lungs, and pancreas. The mesoderm or middle germ layer of the embryo will form the blood, heart, kidney, muscles, and connective tissues. The ectoderm or outermost germ layer of the developing embryo forms epidermis, the brain, and the nervous system.

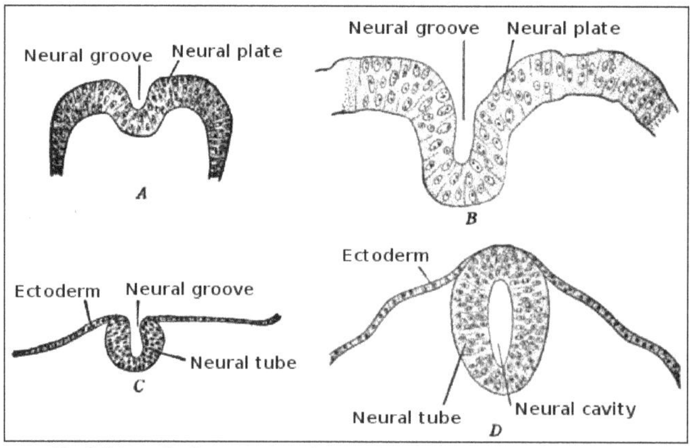

Neural precursor cells fold and elongate to form the neural tube. Mesoderm cells condense to form a rod which will send out signals to redirect the ectoderm cells above. This fold along the neural tube sets up the vertebrate central nervous system.

Mechanism of Organ Formation

While each germ layer forms specific organs, in the 1820's, embryologist Heinz Christian Pander discovered that the germ layers cannot form their respective organs without the cellular interactions from other tissues. In humans, internal organs begin to develop within 3–8 weeks after fertilization. The germ layers form organs by three processes: folds, splits, and condensation. Folds form in the germinal sheet of cells and usually form an enclosed tube which you can see in the development of vertebrates' neural tube. Splits or pockets may form in the germinal sheet of cells forming vesicles or elongations. The lungs and glands of the organism may develop this way.

A primary step in organogenesis for chordates is the development of the notochord, which induces the formation of the neural plate, and ultimately the neural tube in vertebrate development. The development of the neural tube will give rise to the brain and spinal cord. Vertebrates develop a neural crest that differentiates into many structures, including bones, muscles, and components of the central nervous system. Differentiation of the ectoderm into the neural crest, neural tube, and surface ectoderm is sometimes referred to as neurulation and the embryo in this phase is the neurula. The coelom of the body forms from a split of the mesoderm along the somite axis.

Plant Organogenesis

In plants, organogenesis, which is simply the process of forming new organs, occurs continuously and only stops when the plant dies. In the shoot, the shoot apical meristems regularly produce new lateral organs (leaves or flowers) and lateral branches. In the root, new lateral roots form from weakly differentiated internal tissue (e.g. the xylem-pole pericycle in the model plant Arabidopsis thaliana). In vitro and in response to specific cocktails of hormones (mainly auxins and cytokinins), most plant tissues can de-differentiate and form a mass of dividing totipotent stem cells called a callus. Organogenesis can then occur from those cells. The type of organ that is formed depends on the relative concentrations of the hormones in the medium. Plant organogenesis can be induced in tissue culture and used to regenerate plants.

Cellular Differentiation

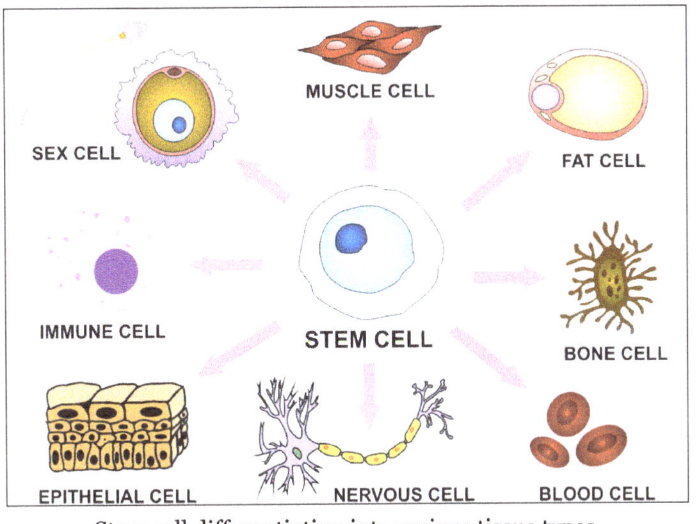

Stem cell differentiation into various tissue types.

Cellular differentiation is the process where a cell changes from one cell type to another. Usually, the cell changes to a more specialized type. Differentiation occurs numerous times during the development of a multicellular organism as it changes from a simple zygote to a complex system of tissues and cell types. Differentiation continues in adulthood as adult stem cells divide and create fully differentiated daughter cells during tissue repair and during normal cell turnover. Some differentiation occurs in response to antigen exposure. Differentiation dramatically changes a cell's size, shape, membrane potential, metabolic activity, and responsiveness to signals. These changes are largely due to highly controlled modifications in gene expression and are the study of epigenetics. With a few exceptions, cellular differentiation almost never involves a change in the DNA sequence itself. Thus, different cells can have very different physical characteristics despite having the same genome.

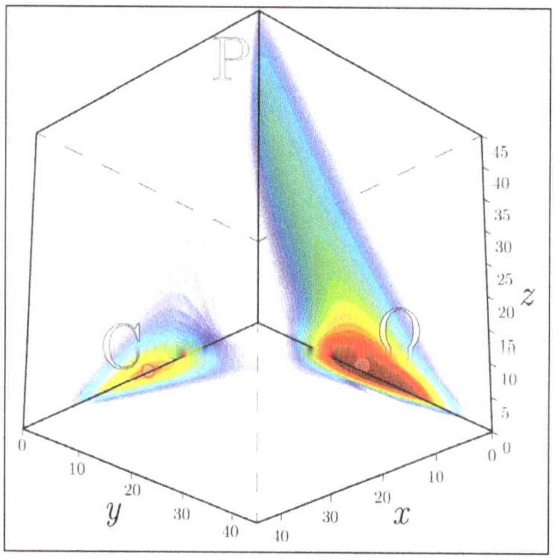

Cell-count distribution featuring cellular differentiation for three types of cells (progenitor z,osteoblast y, and chondrocyte x) exposed to pro-osteoblast stimulus.

A specialized type of differentiation, known as 'terminal differentiation', is of importance in some tissues, for example vertebrate nervous system, striated muscle, epidermis and gut. During terminal differentiation, a precursor cell formerly capable of cell division, permanently leaves the cell cycle, dismantles the cell cycle machinery and often expresses a range of genes characteristic of the cell's final function (e.g. myosin and actin for a muscle cell). Differentiation may continue to occur after terminal differentiation if the capacity and functions of the cell undergo further changes.

Among dividing cells, there are multiple levels of cell potency, the cell's ability to differentiate into other cell types. A greater potency indicates a larger number of cell types that can be derived. A cell that can differentiate into all cell types, including the placental tissue, is known as totipotent. In mammals, only the zygote and subsequent blastomeres are totipotent, while in plants, many differentiated cells can become totipotent with simple laboratory techniques. A cell that can differentiate into all cell types of the adult organism is known as pluripotent. Such cells are called meristematic cells in higher plants and embryonic stem cells in animals, though some groups report the presence of adult pluripotent cells. Virally induced expression of four transcription factors Oct_4, Sox_2, c-Myc, and Klf4 (Yamanaka factors) is sufficient to create pluripotent (iPS) cells from adult fibroblasts. A multipotent cell is one that can differentiate into multiple different, but closely related cell types. Oligopotent cells are more restricted than multipotent, but can still differentiate into a few closely related cell types. Finally, unipotent cells can differentiate into only one cell type, but are capable of self-renewal. In cytopathology, the level of cellular differentiation is used as a measure of cancer progression. "Grade" is a marker of how differentiated a cell in a tumor is.

Mammalian Cell Types

Three basic categories of cells make up the mammalian body: germ cells, somatic cells, and stem cells. Each of the approximately 37.2 trillion ($3.72x10^{13}$) cells in an adult human has its own copy or copies of the genome except certain cell types, such as red blood cells, that lack nuclei in their fully differentiated state. Most cells are diploid; they have two copies of each chromosome. Such cells, called somatic cells, make up most of the human body, such as skin and muscle cells. Cells differentiate to specialize for different functions.

Germ line cells are any line of cells that give rise to gametes—eggs and sperm—and thus are continuous through the generations. Stem cells, on the other hand, have the ability to divide for indefinite periods and to give rise to specialized cells. They are best described in the context of normal human development.

Development begins when a sperm fertilizes an egg and creates a single cell that has the potential to form an entire organism. In the first hours after fertilization, this cell divides into identical cells. In humans, approximately four days after fertilization and after several cycles of cell division, these cells begin to specialize, forming a hollow sphere of cells, called a blastocyst. The blastocyst has an outer layer of cells, and inside this hollow sphere, there is a cluster of cells called the inner cell mass. The cells of the inner cell mass go on to form virtually all of the tissues of the human body. Although the cells of the inner cell mass can form virtually every type of cell found in the human body, they cannot form an organism. These cells are referred to as pluripotent.

Pluripotent stem cells undergo further specialization into multipotent progenitor cells that then give rise to functional cells. Examples of stem and progenitor cells include:

- Radial glial cells (embryonic neural stem cells) that give rise to excitatory neurons in the fetal brain through the process of neurogenesis.
- Hematopoietic stem cells (adult stem cells) from the bone marrow that give rise to red blood cells, white blood cells, and platelets.
- Mesenchymal stem cells (adult stem cells) from the bone marrow that give rise to stromal cells, fat cells, and types of bone cells.
- Epithelial stem cells (progenitor cells) that give rise to the various types of skin cells.
- Muscle satellite cells (progenitor cells) that contribute to differentiated muscle tissue.

A pathway that is guided by the cell adhesion molecules consisting of four amino acids, arginine, glycine, asparagine, and serine, is created as the cellular blastomere differentiates from the single-layered blastula to the three primary layers of germ cells in mammals, namely the ectoderm, mesoderm and endoderm (listed from most distal (exterior) to proximal (interior)). The ectoderm ends up forming the skin and the nervous system, the mesoderm forms the bones and muscular tissue, and the endoderm forms the internal organ tissues.

Dedifferentiation

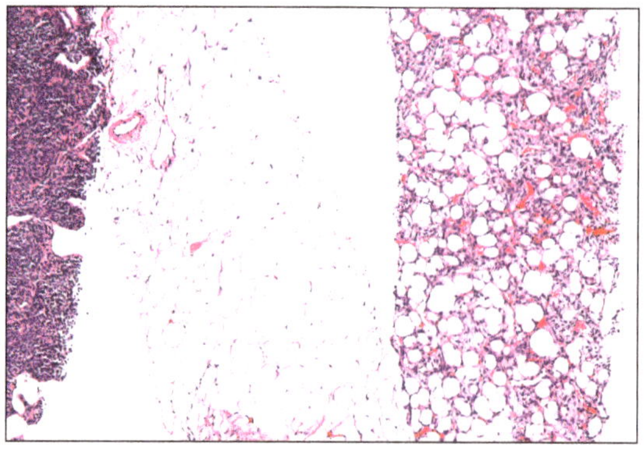

Micrograph of a liposarcoma with some dedifferentiation, that is not identifiable as a liposarcoma, (left edge of image) and a differentiated component (with lipoblasts and increased vascularity (right of image)). Fully differentiated (morphologically benign) adipose tissue (center of the image) has few blood vessels.

Dedifferentiation, or integration is a cellular process often seen in more basal life forms such as worms and amphibians in which a partially or terminally differentiated cell reverts to an earlier developmental stage, usually as part of a regenerative process. Dedifferentiation also occurs in plants. Cells in cell culture can lose properties they originally had, such as protein expression, or change shape. This process is also termed dedifferentiation.

Some believe dedifferentiation is an aberration of the normal development cycle that results in cancer, whereas others believe it to be a natural part of the immune response lost by humans at some point as a result of evolution.

A small molecule dubbed reversine, a purine analog, has been discovered that has proven to induce dedifferentiation in myotubes. These dedifferentiated cells could then redifferentiate into osteoblasts and adipocytes.

Mechanisms

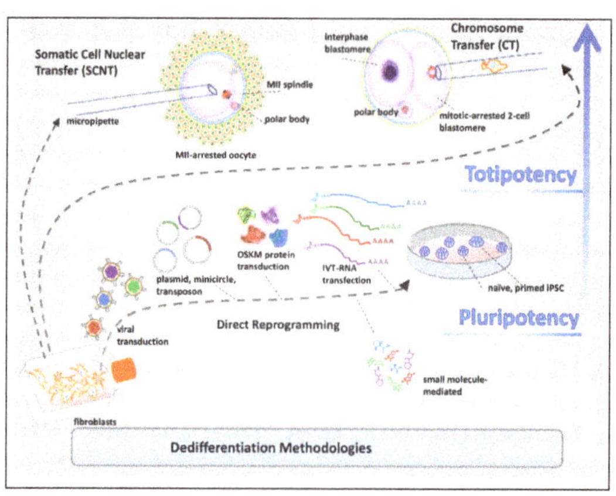

Diagram exposing several methods used to revert adult somatic cells to totipotency or pluripotency.

Each specialized cell type in an organism expresses a subset of all the genes that constitute the genome of that species. Each cell type is defined by its particular pattern of regulated gene expression. Cell differentiation is thus a transition of a cell from one cell type to another and it involves a switch from one pattern of gene expression to another. Cellular differentiation during development can be understood as the result of a gene regulatory network. A regulatory gene and its cis-regulatory modules are nodes in a gene regulatory network; they receive input and create output elsewhere in the network. The systems biology approach to developmental biology emphasizes the importance of investigating how developmental mechanisms interact to produce predictable patterns (morphogenesis). However, an alternative view has been proposed recently. Based on stochastic gene expression, cellular differentiation is the result of a Darwinian selective process occurring among cells. In this frame, protein and gene networks are the result of cellular processes and not their cause.

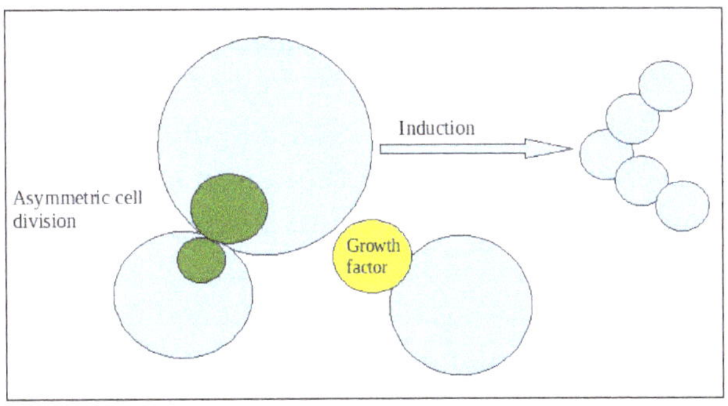

Mechanisms of cellular differentiation.

A few evolutionarily conserved types of molecular processes are often involved in the cellular mechanisms that control these switches. The major types of molecular processes that control cellular differentiation involve cell signaling. Many of the signal molecules that convey information from cell to cell during the control of cellular differentiation are called growth factors. Although the details of specific signal transduction pathways vary, these pathways often share the following general steps. A ligand produced by one cell binds to a receptor in the extracellular region of another cell, inducing a conformational change in the receptor. The shape of the cytoplasmic domain of the receptor changes, and the receptor acquires enzymatic activity. The receptor then catalyzes reactions that phosphorylate other proteins, activating them. A cascade of phosphorylation reactions eventually activates a dormant transcription factor or cytoskeletal protein, thus contributing to the differentiation process in the target cell. Cells and tissues can vary in competence, their ability to respond to external signals.

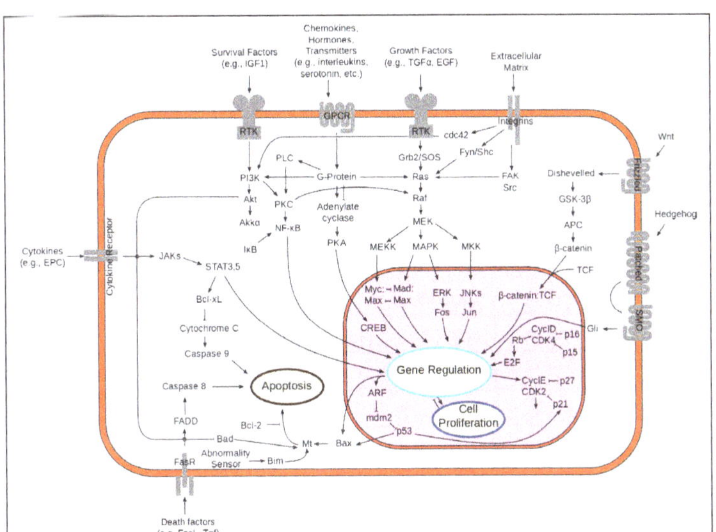

An overview of major signal transduction pathways.

Signal induction refers to cascades of signaling events, during which a cell or tissue signals to another cell or tissue to influence its developmental fate. Yamamoto and Jeffery investigated the role of the lens in eye formation in cave- and surface-dwelling fish, a striking example of induction. Through reciprocal transplants, Yamamoto and Jeffery found that the lens vesicle of surface fish can induce other parts of the eye to develop in cave- and surface-dwelling fish, while the lens vesicle of the cave-dwelling fish cannot.

Other important mechanisms fall under the category of asymmetric cell divisions, divisions that give rise to daughter cells with distinct developmental fates. Asymmetric cell divisions can occur because of asymmetrically expressed maternal cytoplasmic determinants or because of signaling. In the former mechanism, distinct daughter cells are created during cytokinesis because of an uneven distribution of regulatory molecules in the parent cell; the distinct cytoplasm that each daughter cell inherits results in a distinct pattern of differentiation for each daughter cell. A well-studied example of pattern formation by asymmetric divisions is body axis patterning in Drosophila. RNA molecules are an important type of intracellular differentiation control signal. The molecular and genetic basis of asymmetric cell divisions has also been studied in green algae of the genus *Volvox*, a model system for studying how unicellular organisms can evolve into multicellular organisms. In *Volvox carteri*, the 16 cells in the anterior hemisphere of a 32-cell embryo divide asymmetrically, each producing one large and one small daughter cell. The size of the cell at the end of all cell divisions determines whether it becomes a specialized germ or somatic cell.

Epigenetic Control

Since each cell, regardless of cell type, possesses the same genome, determination of cell type must occur at the level of gene expression. While the regulation of gene expression can occur through cis- and trans-regulatory elements including a gene's promoter and enhancers, the problem arises as to how this expression pattern is maintained over numerous generations of cell division. As it turns out, epigenetic processes play a crucial role in regulating the decision to adopt a stem, progenitor, or mature cell fate.

In systems biology and mathematical modeling of gene regulatory networks, cell-fate determination is predicted to exhibit certain dynamics, such as attractor-convergence (the attractor can be an equilibrium point, limit cycle or strange attractor) or oscillatory.

Importance of Epigenetic Control

The first question that can be asked is the extent and complexity of the role of epigenetic processes in the determination of cell fate. A clear answer to this question can be seen in the 2011 paper by Lister R, *et al.*on aberrant epigenomic programming in human induced pluripotent stem cells. As induced pluripotent stem cells (iPSCs) are thought to mimic embryonic stem cells in their pluripotent properties, few epigenetic differences should exist between them. To test this prediction, the authors conducted whole-genome profiling of DNA methylation patterns in several human embryonic stem cell (ESC), iPSC, and progenitor cell lines.

Female adipose cells, lung fibroblasts, and foreskin fibroblasts were reprogrammed into induced pluripotent state with the OCT4, SOX2, KLF4, and MYC genes. Patterns of DNA methylation in ESCs, iPSCs, somatic cells were compared. Lister R, *et al.* observed significant resemblance in methylation levels between embryonic and induced pluripotent cells. Around 80% of CG dinucleotides in ESCs and iPSCs were methylated, the same was true of only 60% of CG dinucleotides in somatic cells. In addition, somatic cells possessed minimal levels of cytosine methylation in non-CG dinucleotides, while induced pluripotent cells possessed similar levels of methylation as embryonic stem cells, between 0.5 and 1.5%. Thus, consistent with their respective transcriptional activities, DNA methylation patterns, at least on the genomic level, are similar between ESCs and iPSCs.

However, upon examining methylation patterns more closely, the authors discovered 1175 regions of differential CG dinucleotide methylation between at least one ES or iPS cell line. By comparing these regions of differential methylation with regions of cytosine methylation in the original somatic cells, 44-49% of differentially methylated regions reflected methylation patterns of the respective progenitor somatic cells, while 51-56% of these regions were dissimilar to both the progenitor and embryonic cell lines. In vitro-induced differentiation of iPSC lines saw transmission of 88% and 46% of hyper and hypo-methylated differentially methylated regions, respectively.

Two conclusions are readily apparent from this study. First, epigenetic processes are heavily involved in cell fate determination, as seen from the similar levels of cytosine methylation between induced pluripotent and embryonic stem cells, consistent with their respective patterns of transcription. Second, the mechanisms of de-differentiation (and by extension, differentiation) are very complex and cannot be easily duplicated, as seen by the significant number of differentially methylated regions between ES and iPS cell lines. Now that these two points have been established, we can examine some of the epigenetic mechanisms that are thought to regulate cellular differentiation.

Mechanisms of Epigenetic Regulation

Pioneering Factors (Oct4, Sox2, Nanog)

Three transcription factors, OCT4, SOX2, and NANOG – the first two of which are used in induced pluripotent stem cell (iPSC) reprogramming, along with Klf4 and c-Myc – are highly expressed in undifferentiated embryonic stem cells and are necessary for the maintenance of their pluripotency. It is thought that they achieve this through alterations in chromatin structure, such as histone modification and DNA methylation, to restrict or permit the transcription of target genes. While highly expressed, their levels require a precise balance to maintain pluripotency, perturbation of which will promote differentiation towards different lineages based on how the gene expression levels change. Differential regulation of Oct-4 and SOX2 levels have been shown to precede germ layer fate selection. Increased levels of Oct4 and decreased levels of Sox2 promote a mesendodermal fate, with Oct4 actively suppressing genes associated with a neural ectodermal fate. Similarly, Increased levels of Sox2 and decreased levels of Oct4 promote differentiation towards a neural ectodermal fate, with Sox2 inhibiting differentiation towards a mesendodermal fate. Regardless of the lineage cells differentiate down, suppression of NANOG has been identified as a necessary prerequisite for differentiation.

Polycomb Repressive Complex (PRC2)

In the realm of gene silencing, Polycomb repressive complex 2, one of two classes of the Polycomb group (PcG) family of proteins, catalyzes the di- and tri-methylation of histone H3 lysine

27 (H3K27me2/me3). By binding to the H3K27me2/3-tagged nucleosome, PRC1 (also a complex of PcG family proteins) catalyzes the mono-ubiquitinylation of histone H2A at lysine 119 (H2AK119Ub1), blocking RNA polymerase II activity and resulting in transcriptional suppression. PcG knockout ES cells do not differentiate efficiently into the three germ layers, and deletion of the PRC1 and PRC2 genes leads to increased expression of lineage-affiliated genes and unscheduled differentiation. Presumably, PcG complexes are responsible for transcriptionally repressing differentiation and development-promoting genes.

Trithorax Group Proteins (TrxG)

Alternately, upon receiving differentiation signals, PcG proteins are recruited to promoters of pluripotency transcription factors. PcG-deficient ES cells can begin differentiation but cannot maintain the differentiated phenotype. Simultaneously, differentiation and development-promoting genes are activated by Trithorax group (TrxG) chromatin regulators and lose their repression. TrxG proteins are recruited at regions of high transcriptional activity, where they catalyze the trimethylation of histone H3 lysine 4 (H3K4me3) and promote gene activation through histone acetylation. PcG and TrxG complexes engage in direct competition and are thought to be functionally antagonistic, creating at differentiation and development-promoting loci what is termed a "bivalent domain" and rendering these genes sensitive to rapid induction or repression.

DNA Methylation

Regulation of gene expression is further achieved through DNA methylation, in which the DNA methyltransferase-mediated methylation of cytosine residues in CpG dinucleotides maintains heritable repression by controlling DNA accessibility. The majority of CpG sites in embryonic stem cells are unmethylated and appear to be associated with H3K4me3-carrying nucleosomes. Upon differentiation, a small number of genes, including OCT4 and NANOG, are methylated and their promoters repressed to prevent their further expression. Consistently, DNA methylation-deficient embryonic stem cells rapidly enter apoptosis upon in vitro differentiation.

Nucleosome Positioning

While the DNA sequence of most cells of an organism is the same, the binding patterns of transcription factors and the corresponding gene expression patterns are different. To a large extent, differences in transcription factor binding are determined by the chromatin accessibility of their binding sites through histone modification and/or pioneer factors. In particular, it is important to know whether a nucleosome is covering a given genomic binding site or not. This can be determined using a chromatin immunoprecipitation (ChIP) assay.

Histone Acetylation and Methylation

DNA-nucleosome interactions are characterized by two states: either tightly bound by nucleosomes and transcriptionally inactive, called heterochromatin, or loosely bound and usually, but not always, transcriptionally active, called euchromatin. The epigenetic processes of histone methylation and acetylation, and their inverses demethylation and deacetylation primarily account for these changes. The effects of acetylation and deacetylation are more predictable. An acetyl group is either added to or removed from the positively charged Lysine residues in histones by enzymes called histone acetyltransferases or histone deacteylases, respectively. The acetyl group prevents Lysine's association with the negatively charged DNA backbone. Methylation is not as straightforward, as neither methylation nor demethylation consistently correlate with either gene activation or repression. However, certain methylations have been repeatedly shown to either activate or repress genes. The trimethylation of lysine 4 on histone 3 (H3K4Me3) is associated with gene activation, whereas trimethylation of lysine 27 on histone 3 represses genes.

In Stem Cells

During differentiation, stem cells change their gene expression profiles. Recent studies have implicated a role for nucleosome positioning and histone modifications during this process. There are two components of this process: turning off the expression of embryonic stem cell (ESC) genes, and the activation of cell fate genes. Lysine specific demethylase 1 (KDM1A) is thought to prevent the use of enhancer regions of pluripotency genes, thereby inhibiting their transcription. It interacts with Mi-2/NuRD complex (nucleosome remodelling and histone deacetylase) complex, giving an instance where methylation and acetylation are not discrete and mutually exclusive, but intertwined processes.

Role of Signaling in Epigenetic Control

A final question to ask concerns the role of cell signaling in influencing the epigenetic processes governing differentiation. Such a role should exist, as it would be reasonable to think that extrinsic signaling can lead to epigenetic remodeling, just as it can lead to changes in gene expression through the activation or repression of different transcription factors. Little direct data is available concerning the specific signals that influence the epigenome, and the majority of current knowledge about the subject consists of speculations on plausible candidate regulators of epigenetic remodeling. We will first discuss several major candidates thought to be involved in the induction and maintenance of both embryonic stem cells and their differentiated progeny, and then turn to one example of specific signaling pathways in which more direct evidence exists for its role in epigenetic change.

The first major candidate is Wnt signaling pathway. The Wnt pathway is involved in all stages of differentiation, and the ligand Wnt3a can substitute for the overexpression of c-Myc in the generation of induced pluripotent stem cells. On the other hand, disruption of ß-catenin, a component of the Wnt signaling pathway, leads to decreased proliferation of neural progenitors.

Growth factors comprise the second major set of candidates of epigenetic regulators of cellular differentiation. These morphogens are crucial for development, and include bone morphogenetic proteins, transforming growth factors (TGFs), and fibroblast growth factors (FGFs). TGFs and FGFs have been shown to sustain expression of OCT4, SOX2, and NANOG by downstream signaling to Smad proteins. Depletion of growth factors promotes the differentiation of ESCs, while genes with bivalent chromatin can become either more restrictive or permissive in their transcription.

Several other signaling pathways are also considered to be primary candidates. Cytokine leukemia inhibitory factors are associated with the maintenance of mouse ESCs in an undifferentiated state.

This is achieved through its activation of the Jak-STAT3 pathway, which has been shown to be necessary and sufficient towards maintaining mouse ESC pluripotency. Retinoic acid can induce differentiation of human and mouse ESCs, and Notch signaling is involved in the proliferation and self-renewal of stem cells. Finally, Sonic hedgehog, in addition to its role as a morphogen, promotes embryonic stem cell differentiation and the self-renewal of somatic stem cells.

The problem, of course, is that the candidacy of these signaling pathways was inferred primarily on the basis of their role in development and cellular differentiation. While epigenetic regulation is necessary for driving cellular differentiation, they are certainly not sufficient for this process. Direct modulation of gene expression through modification of transcription factors plays a key role that must be distinguished from heritable epigenetic changes that can persist even in the absence of the original environmental signals. Only a few examples of signaling pathways leading to epigenetic changes that alter cell fate currently exist, and we will focus on one of them.

Expression of Shh (Sonic hedgehog) upregulates the production of BMI1, a component of the PcG complex that recognizes H3K27me3. This occurs in a Gli-dependent manner, as Gli1 and Gli2 are downstream effectors of the Hedgehog signaling pathway. In culture, Bmi1 mediates the Hedgehog pathway's ability to promote human mammary stem cell self-renewal. In both humans and mice, researchers showed Bmi1 to be highly expressed in proliferating immature cerebellar granule cell precursors. When Bmi1 was knocked out in mice, impaired cerebellar development resulted, leading to significant reductions in postnatal brain mass along with abnormalities in motor control and behavior. A separate study showed a significant decrease in neural stem cell proliferation along with increased astrocyte proliferation in Bmi null mice.

A alternative model of cellular differentiation during embryogenesis is that positional information is based on mechanical signalling by the cytoskeleton using Embryonic differentiation waves. The mechanical signal is then epigenetically transduced via signal transduction systems (of which specific molecules such as Wnt are part) to result in differential gene expression.

In summary, the role of signaling in the epigenetic control of cell fate in mammals is largely unknown, but distinct examples exist that indicate the likely existence of further such mechanisms.

Effect of Matrix Elasticity

In order to fulfill the purpose of regenerating a variety of tissues, adult stems are known to migrate from their niches, adhere to new extracellular matrices (ECM) and differentiate. The ductility of these microenvironments is unique to different tissue types. The ECM surrounding brain, muscle and bone tissues range from soft to stiff. The transduction of the stem cells into these cells types is not directed solely by chemokine cues and cell to cell signaling. The elasticity of the microenvironment can also affect the differentiation of mesenchymal stem cells (MSCs which originate in bone marrow.) When MSCs are placed on substrates of the same stiffness as brain, muscle and bone ECM, the MSCs take on properties of those respective cell types. Matrix sensing requires the cell to pull against the matrix at focal adhesions, which triggers a cellular mechano-transducer to generate a signal to be informed what force is needed to deform the matrix. To determine the key players in matrix-elasticity-driven lineage specification in MSCs, different matrix microenvironments were mimicked. From these experiments, it was concluded that focal adhesions of the MSCs were the cellular mechano-transducer sensing the differences of the matrix elasticity. The non-muscle myosin IIa-c isoforms generates the forces in the cell that lead to signaling of early commitment markers. Nonmuscle myosin IIa generates the least force increasing to non-muscle myosin IIc. There are also factors in the cell that inhibit non-muscle myosin II, such as blebbistatin. This makes the cell effectively blind to the surrounding matrix. Researchers have obtained some success in inducing stem cell-like properties in HEK 239 cells by providing a soft matrix without the use of diffusing factors. The stem-cell properties appear to be linked to tension in the cells' actin network. One identified mechanism for matrix-induced differentiation is tension-induced proteins, which remodel chromatin in response to mechanical stretch. The RhoA pathway is also implicated in this process.

Developmental Biology of the Pancreas

Pancreatic development represents a fascinating process in which two morphologically distinct tissue types must derive from one simple epithelium. These two tissue types, exocrine (including acinar cells, centro-acinar cells, and ducts) and endocrine cells serve disparate functions, and have entirely different morphology. In addition, the endocrine tissue must become disconnected from the epithelial lining during its development.

The adult pancreas is comprised of at least 10 different cells types including those with endocrine function (α, β, δ, γ, and ε), exocrine function (acinar and duct cells), vascular cells, neurons, and mesenchymal cells that are activated in response to injury. The endocrine and exocrine cells are derived from a common endodermal population of cells that also give rise to the liver and parts of the stomach and small intestine. The process by which these cells are specified to take on mature pancreatic phenotypes, and the extrinsic and intrinsic factors that guide these processes.

Morphological Development – from Endoderm to Definitive Pancreas

Specification of the mouse pancreas is initiated from the distal part of the endodermal foregut and proximal part of the midgut beginning at embryonic day (E) 8.5 based on the initial expression of pancreas-specific genes. By E9, the first morphological sign of pancreatic development is the condensation of mesenchyme over the endoderm that will give rise to the dorsal pancreatic bud, distal to where the stomach will form. In humans, this occurs just prior to 26 days of gestation. At E9.5, a clustering or anlage of cells emerges from the dorsal aspect of the gut tube at the point where the notochord comes in contact with the gut. This relationship between the notochord and endoderm is of critical importance. Shortly after, the paired dorsal aorta move toward the midline and fuse, thereby displacing the notochord from the developing pancreas. Less than one day later in mice, or 6 days later in humans, a second ventrally located anlage appears, just distal to the developing liver, extending from the common bile duct. The two buds continue to develop independently, affected by different surrounding tissues, until approximately E13 (days 37 to 42 in humans), at which point, the rotational movement of the developing gut tube and elongation of both buds causes alignment and fusion into a single organ. In the majority of cases, one of the pancreatic ducts becomes patent, leaving a single common pancreatic duct, the Duct of Wirsung, for the organ. In about 10% of people, the two buds do not fuse leading to pancreatic divisum and the presence of a second duct, termed the Duct of Santorini or accessory duct. Pancreatic divisum is the most common pancreatic developmental anomaly, and in some cases, is an increased risk for recurrent pancreatitis. By E15, clear acinar cell clusters and presumptive islets are apparent, and differentiation markers for all pancreatic cell types are expressed.

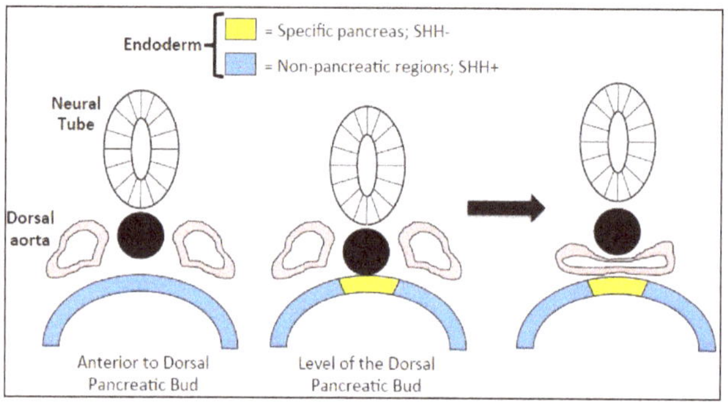

Specification of the dorsal pancreatic bud.

In figure, foregut endoderm anterior to the putative pancreatic bud expresses sonic hedgehog (SHH) which does not allow for pancreatic differentiation. Interactions with the notochord (filled black circle) lead to repression of SHH and progression to a pancreatic fate. This fate is maintained even after the paired dorsal aorta fuses at the midline displacing the notochord from direct interaction with the endoderm.

In general, pancreatic development can be separated into three time periods. The primary transition in mice encompasses E9-12 when specification of different pancreatic cell types takes place. During the primary transition, the pancreatic buds initially appear as stratified epithelium with a centrally located lumen. Cells within the anlage maintain multipotency, with the ability to give rise to endocrine, acinar and duct cells. The number of multipotent pancreatic progenitor cells (MPCs) is determined during the primary transition and appears to dictate the final organ size. Beginning at E10.75, microlumen structures begin to form, with cells establishing an apical polarity eventually leading to duct formation. Integrin-based interactions between outer cells of the developing buds and the adjacent basement membrane are required for initiation of branching. Minimal expression of pancreatic hormones and digestive enzymes is observed, although endocrine cells expressing multiple hormones exist, present as clusters within the dorsal bud. Whether these initial endocrine cells are maintained in the adult is still debatable.

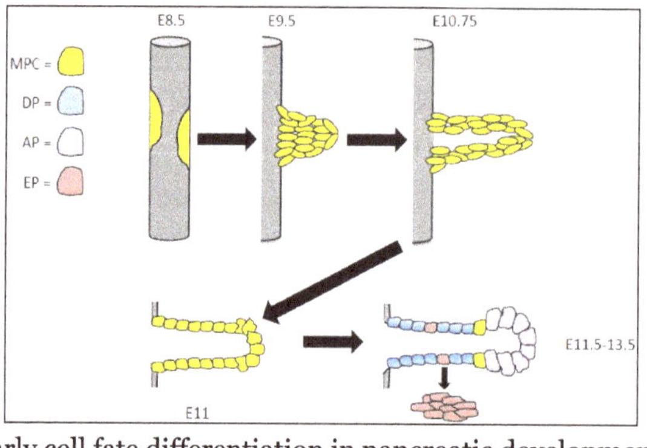

Early cell fate differentiation in pancreatic development.

In figure, Multipotent pancreatic progenitor cells (MPCs) are first identified at embryonic day (E) 8.5. These cells maintain multipotency through to E10.5, at which point they begin to differentiate into acinar progenitor (AP), endocrine progenitor (EP) and duct progenitor (DP) cells. MPCs are maintained at the neck of pancreatic buds until E14.5.

The secondary transition occurs between E12 and E15 and is characterized by significant expansion and branching of the pancreatic bud, delamination of individual presumptive islet cells from the pancreatic epithelium, and enhanced expression of both hormones and digestive enzymes. At this point, most of the cells have been specified to become either endocrine, acinar or duct cells, with only a small proportion of cells maintaining multipotency. The third stage of pancreatic development occurs after E15 and extends postnatally with continued maturation and expansion of the pancreas. While controversial, it appears that all cells within the pancreas after E15 are unipotent, determined to be either acinar, duct or endocrine in nature. However, multipotency can be stimulated in the adult by injury.

High-powered resolution of the developing pancreas reveals a relationship between cell location within the branching epithelium and the different cell types. Progenitor cells giving rise specifically to acinar cells are located predominantly at the tips of the branching duct network. Conversely, mature duct and endocrine cells are derived from the trunks of these branches. MPCs exist at the transition position between the tips and the trunks. Indeed, the location of these cells provided support for the hypothesis that centroacinar cells, which sit at the junction between acini and ducts, are pancreatic stem cells in the adult. However, studies examining centroacionar cells as stem cells have have not supported such a role. As mentioned above, by E15, MPCs are rare. Evidence suggests that presumptive endocrine cells undergo a mesenchymal-epithelial transition process and coalesce to form the primitive islets of Langerhans. Further recruitment, proliferation and maturation give rise to the mature islets postnatally. In mice, the islets consist of an inner core of β cells, which secrete insulin, surrounded by a mantel of α (glucagon), δ (somatostatin), ε (ghrelin) and PP (pancreatic polypeptide) cells. In humans, these cells are positioned throughout the islets.

Differentiation of acinar cells is less overt. Acinar cells maintain attachment to the duct network and develop a classical epithelial phenotype based on their polarity and intercellular junctions. Expression of carboxypeptidase (CPA) is observed as early as E9, but CPA expression can be observed at early time points in MPCs as well as in acinar progenitor cells. More extensive expression of pancreatic enzymes begins around E12.5, and zymogen granules, the morphological hallmark of acinar cells, are first observed around E16.5. Acinar cells do exhibit polarity prior to birth based on the localization of tight and adherens junctions. However, the high degree of organization within acinar cells is first observed following birth. At this point, mature acinar cell morphology has been obtained, while further proliferation and maturation at the molecular level continues to occur until after weaning, presumably due to alterations in dietary makeup.

The three main pancreatic cell types are derived from a common progenitor pool.

Elegant genetic lineage tracing experiments in mice have defined the progression of MPCs to individual unipotential populations as well as identified specific markers for this process. Mouse models were developed in which cre recombinase expression was driven in specific cell populations during development. Initially, these experiments involved spatially expressing cre recombinase based solely on the promoter used and mating these "driver" mice to those containing a loxP-stop-loxP (LSL) site upstream of a reporter gene such as green fluorescent protein. Since this LSL site is downstream of a constitutively active and ubiquitous promoter, any cells derived from the initial expressing cells would be positive for the reporter gene. A number of groups used a system in which the cre-recombinase could be manipulated temporally through the addition of a tamoxifen-inducible tag linked to cre. First, prior to E12, MPCs exist that give rise to all cell types. Second, acinar progenitor cells are distinct from endocrine/duct progenitor cells indicating that duct cells have a closer developmental origin to endocrine cells. Third, MPCs do not normally exist within the adult, but can be stimulated to re-appear by injury to the pancreas. Fourth, acinar cells have the ability of de-differentiation into duct-like cells, and likely contribute to initial PanIN formation. These studies were also instrumental in defining that MPCs are located at the junction between tip and trunk cells of the branching pancreatic duct and can be identified by a specific combination of transcription factors (TFs). MPCs identified during the primary transition express Pancreatic Transcription Factor 1A (PTF1A), Pancreatic-Duodenal Homeobox 1 (PDX1), Sry-related HMG box (SOX9) and NK Homeobox 6.1 (NKX6.1). Hepatocyte Nuclear Factor 1β (HNF1β) defines epithelial cells in the trunk that give rise to endocrine and duct cells but not acinar cells, although there is some controversy as Wright's group has suggested the HNF1β may actually be expressed in MPCs as well. Whether this difference reflects the time points examined or differences in analysis is unclear. During the secondary transition, PTF1A and NKX6.1 become mutually exclusive with PTF1A cells specifying an acinar cell phenotype and NKX6.1 found in duct/endocrine precursors.

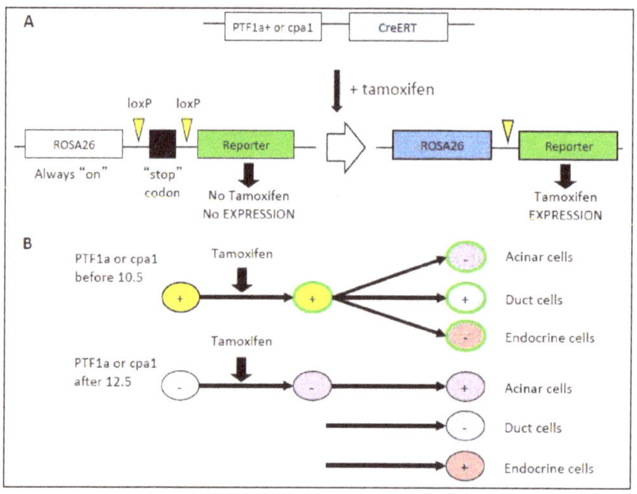

Lineage tracing during pancreatic development.

In figure, (A) Scheme for permanently marking cells during development. Expression of cre recombinase is limited spatially by the promoter, in this case Ptf1a or cpa1(carboxypeptidase 1), and temporally by the addition of a tamoxifen inducible estrogen receptor (ERT) fused to cre. The constitutively active ROSA26 drives expression of the reporter gene, which is prevented by the presence of stop codons flanked by IoxP sites (target for cre recombinase). Upon the introduction of tamoxifen, the stop codons are removed leading to reporter gene expression. (B) Using either the Ptf1a or cpa1 promoter to drive creERT expression, introduction of tamoxifen early in development leads to reporter expression in all pancreatic cell types. Introduction of tamoxifen after E12.5 results in only acinar cells being labeled.

Developmental Relationships that affect Pancreatic Development

Specification of the pancreas is dictated by interactions with a number of structures developing in close proximity to the developing endoderm. As mentioned, the ventral pancreatic duct shares a common cellular origin with the liver and, therefore, signals must be present that dictate one fate from the other. Interestingly, it appears that the dorsal and ventral aspects of the endoderm are governed by either significantly different signaling events, or are able to process specific signaling pathways, such as Fibroblast growth factor (FGF) signaling, in different ways. This suggests that the default differentiation pattern differs for dorsal and ventral endoderm. Indeed, culturing ventral endoderm independent of other structures or signals results in pancreatic differentiation, based on the expression of both endocrine and exocrine cells. This suggests that the default pattern of differentiation for the ventral endoderm is pancreas, and that surrounding tissue in the embryo repress pancreatic development to become other organs. For example, signals from the cardiac mesoderm and the septum transversum, the structure that eventually gives rise to the diaphragm and parts of the ventral mesentery, promote liver differentiation at the expense of the pancreas. Conversely, culturing of the dorsal endoderm does not give rise to pancreas suggesting that signals are required to promote differentiation and the default differentiation program is not pancreas.

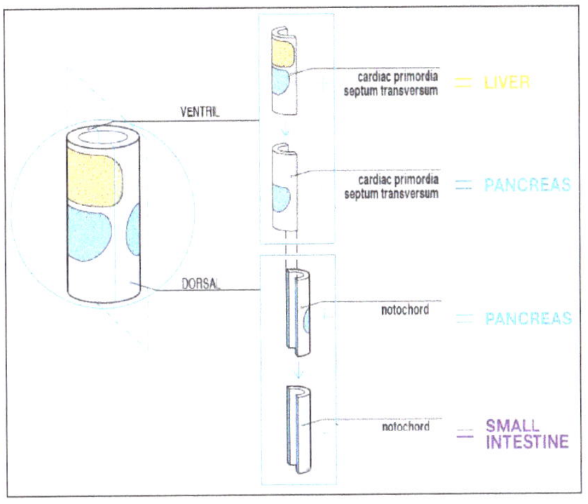

The dorsal and ventral pancreatic buds have different default patterns for development.

In figure, the pancreas is derived from a ventral pancreatic bud (VPB) and a dorsal pancreatic bud (DPB) that first appear around E9.5 in mice. The ventral endoderm gives rise to liver tissue due to inhibitory signals from the septum transversum and cardiac primordia. In the absence of these signals, pancreatic development is the default pattern of differentiation. The dorsal endoderm requires inductive signals from the notochord to promote pancreas development. In the absence of these signals, the dorsal endoderms will differentiation into non-pancreatic structures, such as the small intestinee.

Mesenchyme

Coalescence of the mesenchyme at the level where the dorsal pancreas will form is the first morphological sign of pancreatic development. Removal of the mesoderm, or the fibroblasts within the mesoderm, prior to pancreatic specification results in pancreatic agenesis. Mesoderm removal following specification results in a reduction of the total pancreatic size indicating an ongoing requirement for mesoderm signaling to attain complete organ development. Interestingly, culturing of pancreatic mesenchyme with other sections of the dorsal endoderm can promote pancreatic differentiation, while mesenchyme from other regions of the anterior-posterior axis does not have this ability. This suggests that the mesenchyme provides signals that promote pancreatic specification, yet limits differentiation, thereby allowing expansion of the organ. These signals come in a variety of sources.

Physical interactions between the mesoderm and developing pancreatic bud affect differentiation and maturation in multiple ways. Physical interaction is particularly important in specification of the exocrine pancreas. Glypicans and syndecans promote exocrine cell differentiation, and Notch signaling promotes PTF1A's function while repressing Neurogenin (NGN) 3 expression. PTF1A and NGN3 are key transcriptional regulators that promote exocrine and endocrine cell specification, respectively. Laminin-1, a key extracellular component within the mesenchyme, promotes pancreatic duct formation and substrates that contain laminin-1, such as matrigel, can promote duct formation from early pancreatic buds possibly through α6-containing integrin complexes. Furthermore, direct interaction between the mesenchyme and endoderm represses islet formation. Only after extensive branching excludes the mesenchyme from the developing ductal network can progenitor cells undergo epithelial to mesenchymal transition (EMT) and coalesce to form definitive islets. Finally, direct interactions between the mesenchyme and developing pancreatic bud maintain a progenitor like state. Inhibition of Epi-CAM (epithelial cell adhesion molecules) in pancreatic bud cultures induces premature endocrine cell maturation, suggesting Epi-CAM helps maintain a ductal endocrine progenitor cell type.

In addition to physical interactions, secreted factors from the surrounding mesenchyme have also been identified that either promote or repress pancreatic specification and differentiation. Signaling through the TGFβ superfamily affects many steps in pancreatic differentiation. TGFβ signaling from the mesenchyme promotes endocrine differentiation from MPCs, and TGFβ treatment of pancreatic dorsal bud tissue leads to endocrine differentiation at the expense of exocrine tissue. Activin, another member of the TGFβ superfamily, also promotes endocrine differentiation while blocking branching and proliferation of the pancreatic bud and repressing exocrine differentiation. Inhibition of activin using follistatin, promotes exocrine cell differentiation at the expense of endocrine tissue. Disruption of TGFβ signaling affects the secondary transition and once pancreatic fates have been determined, TGFβ signaling is required to maintain mature acinar cell fate as over-expression of dominant negative TGFβRII in the pancreas leads to acinar to duct cell metaplasia (ADM). Additionally, a number of Fibroblast Growth factors (FGFs) are secreted by the mesenchyme, which affects pancreatic development. FGF4 represses endocrine differentiation and allows for expansion of the exocrine compartment, while FGF7 and FGF10 repress exocrine differentiation, thereby allowing proliferation and expansion of the pancreatic bud. Fgf10-/- mice exhibit pancreatic agenesis due to the inability to repress sonic hedgehog, a repressor of pancreatic differentiation.

Vasculature

At E8.5, de novo blood vessels coalesce from endothelial precursors immediately dorsal to the pre-pancreatic endoderm. Paracrine signaling from these blood vessels is crucial in promoting early pancreatic organ specification, particularly in the dorsal pancreatic bud. The importance of blood flow was first identified in a genetically modified mouse model containing a targeted deletion of Cdh2, the gene encoding N-cadherin. Mice completely null for N-cadherin show dorsal pancreatic agenesis, initially suggested to be due to impaired interactions between the mesenchyme and pancreatic endoderm. However, restoring N-cadherin specifically in the heart rescued both heart and pancreatic development indicating that maintaining blood flow to the pancreatic anlage is required for dorsal pancreatic formation. Further studies showed that signals from the vasculature could be from both the circulating blood as well as the endothelial cells of the dorsal aorta and surrounding mesenchyme.

Interactions between the vasculature and endoderm also limit pancreatic differentiation. Over-expression of VEGF-A with the Pdx1 promoter leads to insulin production within the stomach. Conversely, over-expression of VEGF-A within the pancreatic anlage greatly enhances vasculature development and significantly reduced pancreas formation, specifically affecting exocrine pancreatic mass. Ex vivo cultures in which the developing pancreas was treated with VEGF showed increased vasculature and greatly reduced size of the pancreas, stalling epithelial cells in a progenitor like state. Conversely, ablation of endothelial cells following pancreatic specification leads to more promiscuous differentiation of acinar cells, ultimately reducing the size of the pancreas as well. Therefore, the endothelium plays a dual role in dorsal pancreatic development. During the primary transition, the vasculature provides signals that specify pancreatic differentiation from the endoderm, yet maintain trunk progenitor populations in an undifferentiated state, preventing premature development of specification to an endocrine fate. During the secondary transition, the vasculature restricts exocrine differentiation.

Again, the signals derived from the vasculature are diverse. Showed that direct interaction with the aortic endothelial cells induced pancreatic bud-like structures and PDX1 expression in the adjacent endoderm. Endothelial cells are also required for PTF1A expression in the dorsal foregut. Conversely, development of the ventral pancreatic bud does not appear to require direct endothelial interactions but rather relies on paracrine factors for development. For example, Sphingosine-1-phosphate, found in the blood, promotes recruitment to a pancreatic cell fate, and induces proliferation of pancreatic mesenchymal and endothelial cells.

Notochord and Sonic Hedgehog

A critical negative regulator of pancreatic development is sonic hedgehog (SHH). This is evidenced by experiments in which SHH signaling is elevated either through a transgenic approach where the Pdx1 promoter drives SHH expression or by ablating the Patched receptor (Ptc). Ptc represses smoothen, a downstream mediator of SHH signaling. Both Pdx1-Shh and $Ptc^{-/-}$ mice exhibit pancreatic agenesis. These experiments highlight the importance of repressing SHH at the initiation of pancreatic development. Forcing SHH expression, or the closely related Indian Hedgehog (IHH), after pancreatic specification (E12.5) by using the Pax4 promoter, still leads to a significant loss of exocrine and endocrine tissue that is replaced by mesenchyme and stroma. Therefore, hedgehog suppression must be maintained to ensure proper pancreatic development. Indeed, activation of SHH signaling is observed during pancreatic injury correlating to acinar to duct cell metaplasia. Surprisingly, in zebrafish endoderm, ectopic SHH signaling expands the Pdx1-positive pancreas domain, while inhibiting hedgehog activity results in pancreatic agenesis indicating species differences in signaling pathways.

While the mesenchyme expresses factors that promote SHH repression, including activin and FGFs, the notochord appears to be the primary source of these signals at the initiation of dorsal pancreas induction. The foregut expresses SHH along its length except at the point where the notochord comes in close proximity to the endoderm. Ectopic notochord engraftment represses SHH expression in the lateral and ventral endoderm, and SHH neutralizing antibodies can induce pancreas gene expression in dorsal endoderm explants. However, blocking SHH in non-pancreatic endoderm does not lead to pancreatic development. This suggests that signals from the notochord are not instructive but, rather, permissive. In agreement with this permissive model, $Shh^{-/-}$ mice do not show expansion of the pancreas along the foregut, indicating that additional factors are required.

Ventral Bud Differentiation, Cardiac Mesenchyme and Septum Transversum

The signals derived from the notochord and mesenchyme that permit pancreatic development are activin (TGFβ family) and FGFs. Surprisingly, these same factors repress ventral pancreatic differentiation, highlighting differences between in the developmental processes of the two pancreatic buds. A number of other findings highlight differences between ventral and dorsal pancreatic bud development. $Hhex^{-/-}$ embryos fail to specify a ventral pancreas but the dorsal bud develops without issues. Alternatively, loss of Raldh2, which is required for endogenous retinoic acid signaling, prevents development of the dorsal pancreas while the ventral portion of the organ remains unaffected.

Since ventral pancreatic bud formation is the default pattern of ventral foregut differentiation, the factors that govern differentiation of this anlage initially repress pancreatic fate. Similar to the repression of SHH by FGFs and activin in the dorsal pancreatic bud, these factors repress pancreatic specification, thereby allowing differentiation of the liver. The ventral foregut endoderm lies in close juxtaposition to the cardiac mesoderm and septum transversum mesenchyme, and these tissues induce liver fate at the expense of the pancreas. Bone morphogenetic proteins (BMPs) also promote epigenetic reprogramming of ventral endoderm to a liver fate. In the absence of these structures or signaling molecules, the endodermal cells activate pancreatic differentiation pathways. Once liver development proceeds, the repressive signals are blocked allowing for pancreatic development from the more distal ventral endoderm to occur. In support of such a model, it has been suggested that $Hhex^{-/-}$ fail to develop a ventral pancreas due to lack of proliferation of the endoderm, which would displace the putative pancreatic bud from influences of the cardiac mesoderm. Supporting this hypothesis, $Hhex^{-/-}$ endoderm cultured in vitro is fully capable of expressing PDX1.

While these various developmental structures have a significant impact on pancreatic development, their effects are governed by the permissive nature of the developing endoderm.

Transcription Factors Governing Pancreatic Development

In this topic, we will discuss transcription factors (TFs) which govern tissue and cell development. During pancreatic development, coordinated gene expression is critical for cell specification, expansion of the progenitor population, differentiation and then final maturation and function. The TF network has been well characterized in the pancreas, especially regarding β-cell development, with many TFs exhibiting dual roles during development and adult function. In early development, specific TFs govern the establishment and maintenance of the MPCs, while in mature tissue, TFs affect cell specific function. A complete description of each TF is not possible. Therefore, this topic will focus on identifying TFs linked to factors that define MPCs or are involved in exocrine tissue development as more extensive reviews on TFs that affect endocrine and specifically β-cell development can be found elsewhere.

Transcription factors Involved in Specifying MPCs from Endoderm

In the 1990s, two key TFs - PDX1 and PTF1a - were identified as critical for specification and differentiation of all cell types within the pancreas. Using genetically modified mice, both PDX1 and PTF1a were shown to be absolutely required for differentiation of both endocrine and exocrine pancreatic compartments indicating these factors are required for initial specification of MPCs from the endoderm. Since that time, several other TFs have been linked to specification of MPCs or shown to be required for expansion or maturation of the exocrine pancreas. Many of these proteins exhibit defined roles in development and mature cell function, and have been implicated in various pathologies such as pancreatitis or pancreatic cancer.

Pancreatic and Duodenal Homeobox 1

Pancreatic and duodenal homeobox 1 (PDX; alternatively known as Insulin Promoting Factor 1 (IPF1), or Somatostatin-Transactivating Factor 1(STF1)) is a homeobox containing transcription factor. PDX1 is expressed in the developing mouse as early as E8.5 and maintained in all cells of the pancreas until E17.5, when it becomes restricted to islets. Initial expression of PDX1 appears to delineate a region of the developing gut that later gives rise to the stomach, small intestine and bile duct, suggesting that it is involved in specifying this region of the gut tube to a subset of tissues. Germline deletion of Pdx1 in mice leads to an absence of all pancreatic tissues and neonatal lethality by postnatal day 5 due to hyperglycaemia. Examination of the pancreas during development indicates that initial specification of the gut to become pancreas, as well as initial budding, still occurs in $Pdx1^{-/-}$ mice. However, subsequent expansion and branching of the tissue does not occur, indicating that expansion and complete differentiation of the pancreas requires PDX1 activity. In the mature pancreas, PDX1 was initially described as limited to β and d-cells of the islets, only being re-expressed in acinar cells upon pancreatic injury. However, recent studies have identified low levels of PDX1 in most mature acinar cells, and targeted deletion of Pdx1 in post-developmental acinar cells shows that PDX1 is required for maintaining this compartment of the tissue.

Heterozygous Pdx1 ($Pdx1^{+/-}$) mice initially develop normally. However, over time $Pdx1^{+/-}$ mice develop insulin insufficiency, indicating a requirement for high levels of PDX1 in insulin production. Transient loss of PDX1 in adult tissue using a tet-inducible system supports the ongoing requirement for PDX1 in maintaining β cell function, as loss of PDX1 leads to decreased insulin accumulation and apoptosis. In humans, homozygous mutation of PDX1 results in pancreatic agenesis, while heterozygous mutations of PDX1 are found in patients with Maturity Onset Diabetes of the Young (MODY). Genetic analysis identified the PDX1 gene as MODY4, and PDX1 can directly target and promote the expression of insulin and GLUT2 in β-cells.

Pancreatic Transcription Factor 1A

Pancreatic Transcription Factor 1a (PTF1a; alternatively known as PTF1-p48 or p48) was first identified by its ability to bind combined E-box (CANNTG) and G-box (TGGGA) sequences found in promoters of digestive enzymes trypsin, cpa1, amylase, and elastase. Early in pancreatic development, PTF1a is co-expressed with PDX1 in MPCs. During the secondary transition, PTF1a becomes restricted to acinar progenitor cells. PTF1a is a basic helix-loop-helix (bHLH) transcription factor that forms an unconventional protein trimer with ubiquitously expressed E proteins such as E12 and E47, and RBPJ (Recombinant binding protein J), a downstream mediator of NOTCH signalling. Combined, these factors make up the PTF1 complex, which specifies pancreatic acinar cells. However, while associated with RBPJ, these acinar cell progenitors do not undergo differentiation, thereby allowing for expansion of the progenitor population. Upon commitment to acinar cell differentiation, PTF1a preferentially interacts with the RBPJ-like (RBPJL) protein, and nuclear hormone receptor NR5A2 to drive pancreatic differentiation of acinar progenitor cells.

Targeted deletion of Ptf1a in mice results in postnatal lethality immediately after birth, likely due to neuronal functions, as PTF1a is also expressed in the neural tube and cerebellum. Ptf1a-deficient mice have no pancreatic tissue at birth with PTF1a-deficient cells instead contributing to the common bile duct and duodenum. Interestingly, the result of over-expressing PTF1a depends on developmental state. Forced PTF1 expression early in development in cells that would typically give rise to the duodenum and stomach leads to expansion of the pancreas, with increased differentiation of duct, islet and acinar cell compartments. In ES cells, forced PTF1a expression requires MIST1 to activate expression enzyme gene expression.

ISL1

Islet 1 (Isl1) is a member of the LIM/homeodomain family of transcription factors, expressed in a number of tissues during embryogenesis. In the developing pancreas, ISL1 expression is initially observed within the mesenchyme and later in pancreatic islets. Targeted deletion of Isl1 in mice leads to deficits in pancreatic development and provided initial evidence that the mesenchyme has a role in affecting pancreatic differentiation. Culturing putative mouse Isl1-/- pancreatic endoderm with wild type mesenchyme restored pancreatic differentiation confirming that ISL1 acts as non-cell autonomous signal in early pancreas development. While ISL1 is restricted to pancreatic islets in the adult, expressing Isl1 in acinar cells using an adenoviral-mediated approach, induces tubular complex formation through acinar to ductal metaplasia, in which exogenous expression facilitated this process.

Hepatocyte Nuclear Factors

The Hepatocyte Nuclear Factors (HNFs) are a homeodomain-containing superfamily of TFs that affect multiple stages of development. HNF1b encodes a protein that binds DNA as a homodimer or heterodimer along with HNF1a. During endoderm development, expression of HNF6 activates HNF1b expression. HNF1b is critical for endoderm development with HNF1b+ cells giving rise to liver, kidney, pancreas, bile ducts, urogenital tract, lung, thymus and gut. Mice with a germline deletion of the Hnf1b gene have a very similar phenotype to $Ptf1a^{-/-}$ mice, with no pancreatic development observed at birth. Temporal and spatial deletion of HNF1b throughout pancreatic development has demonstrated important roles in MPC expansion, and acinar, ductal and endocrine cell differentiation. In humans, mutations in the HNF1B gene result in renal cysts, diabetes syndrome and noninsulin-dependent diabetes mellitus. Some individuals with HNF1B mutations display pancreatic exocrine dysfunction, pancreatic hypoplasia, and, in some cases, complete absence of the pancreatic body and tail suggesting specific loss of dorsal pancreatic development.

HNF6 (Hepatic Nuclear Factor 6; also known as ONECUT1 or OC-1) binds DNA to elicit transcription of genes that are important throughout endodermal development. In pancreatic development, HNF6 expression begins prior to PDX1 in MPCs and is critical to several specific developmental functions depending on developmental time point. After the onset of Pdx1 expression, HNF6 is crucial for endocrine cell development. HNF6 induces HNF1B expression, which promotes a wave of proliferation and initiates Neurogenin 3 (Ngn3) expression. Following this increase in Ngn3 expression, HNF6 expression becomes limited to duct cells and is very marginally expressed in acinar cells. Hnf6$^{-/-}$ mice exhibit a hypoplastic pancreas with effects mostly limited to non-acinar cell compartments suggested HNF6 is dispensable for acinar cell differentiation.

Sex-determining Region Y-box (SRY-box) Containing Gene 9 (SOX9)

SOX9 is a member of the SRY-related, high mobility group box transcription factors. Early in development, between E9-E12.5, SOX9 expression overlaps with PDX1 and marks MPCs. SOX9 is required for pancreatic cell fate and repression of intestinal differentiation through binding lineage specific promoters and enhancers. A FGF10/SOX9/FGFR2 feed forward loop maintains pancreatic organ identity in MPCs. While MPCs express SOX9, as cells begin to differentiate SOX9 expression decreases. By E14, SOX9 accumulation is restricted to a subpopulation of PDX1+ cells located within the epithelial cords. Pancreas-specific knockout of Sox9 using the Pdx1-Cre driver leads to stunted growth, elevated blood glucose levels, and death by postnatal day 4. Gross morphological assessment showed hypoplasia of both pancreatic buds beginning at E11.5, due to a reduction of PDX1+ cell proliferation and increase in cell death. Haplo-sufficent Sox9 mice and individuals with Campomelic dysplasia (CD), a semi-lethal skeletal malformation syndrome due to a heterozygous loss-of-function mutation in SOX9, have reduced islet cell mass, but the exocrine tissue remains unaffected. In the adult, SOX9 accumulates in some duct and centroacinar cells, and targeted deletion of Sox9 in the adult pancreas of mice leads to a cystic phenotype and decreased expression of duct markers, indicating SOX9's role in duct cell maintenance.

GATA Transcription Factors

GATA transcription factors are characterized by their ability to bind the DNA consensus site (A/T) GATA(A/G). GATA4 and GATA6 have ~85% homology, containing a highly conserved DNA binding domain consisting of two zinc finger motifs. Initial expression of GATA4 and GATA6 occurs early in development and is associated with the development of many organs of endodermal and mesodermal origin. GATA4 and GATA6 are expressed between E9.5-E15.5 in the developing pancreas. At E9.5, GATA4 and GATA6 are co-expressed throughout the pancreatic epithelium. As development continues, GATA4 accumulation becomes restricted first to the tips of branching ducts and then acinar cells by E16.5 where expression is continually reduced until E18.5 when it is undetectable in exocrine cells. However, in the adult GATA4 is detectable in subsets of α and β cells. GATA6 expression overlaps GATA4 expression until E14.5-E15.5, when it then becomes restricted to the cells of the endocrine pancreas and duct epithelium. Transgenic mouse models in which GATA6 targets are repressed results in pancreatic agenesis. This is not the case when GATA4 targets are repressed suggesting GATA6 is a critical for pancreatic development, whereas GATA4 is dispensable.

Transcription Factors Promoting Acinar Cell Differentiation

Nuclear Receptor 5a2

NR5A2, a member of the orphan nuclear receptor family, is critical for embryogenesis as germ line Nr5a2 deletion results in early embryonic lethality. NR5A2 maintains the expression of embryonic stem cell identity genes such as Oct4 and Nanog, as well as genes involved in self-renewal and growth. Temporal and spatial expression of Nr5a2 determines its function, suggesting that NR5A2 function may link early endoderm development and endoderm differentiation. NR5A2 is expressed in the gastrointestinal endoderm and required for the expansion of pancreatic epithelium. In pancreatic progenitor cells, NR5A2 promotes expression of genes involved in hepatic and pancreatic maturation. In zebrafish, null Nr5a2 mutations have no exocrine pancreas while development of the endocrine pancreas remains largely unaffected. In the mature pancreas, NR5A2 regulates the expression of other protein complexes driving the assembly of digestive enzymes.

Hairy and Enhancer of Split-1

HES1 (hairy and enhancer of split-1) is one of seven Hes gene family members, which encode proteins that heterodimerize with other bHLH proteins, normally leading to target gene repression. HES1 is the key target of NOTCH signalling in pancreatic development. In the absence of NOTCH, Recombining Binding Protein Suppressor of Hairless (RBPJ) inhibits HES1 expression. When NOTCH signalling is activated, HES1 is expressed and represses expression of NOTCH ligands Dll1 and Jagged1 and the endocrine TF, Ngn3. In pancreatic progenitor cells, HES1 expression inhibits the expression of Ptf1a and Ngn3. In Hes1$^{-/-}$ mice, ectopic pancreatic tissue is found in the extrahepatic biliary system, posterior stomach, and proximal duodenum. A conversion of the biliary system into pancreatic tissue has also been noted in Hes1$^{-/-}$ deficient mice, suggesting HES1 restricts pancreatic differentiation from the endoderm. Subsequent studies showed that HES1 and SOX17 are part of a complex regulatory network that controls organ cell fate commitment of pancreatobiliary progenitors in the ventral foregut, by restricting ectopic pancreatic development.

MIST1

MIST1 is a member of the bHLH family of transcription factors expressed at high amounts in mature acinar cells. Activated shortly after PTF1a in a subset of pancreatic cells at E10.5, it is unclear whether MIST1 marks only potential acinar cells or MPCs. By E14.5, MIST1 clearly marks committed acinar cells and is excluded from duct cells through the rest of development and in the adult pancreas. Deletion of Mist1 leads to loss of acinar cell organization and incomplete maturation of acinar cells. While still functional, Mist1$^{-/-}$ cells are more likely to undergo acinar-to-duct cell metaplasia, and these animals are more sensitive to pancreatic injury.

Several targets of MIST1 transcriptional activity, including Cx32, Rab3d and Atp2c2, have been identified. However, it is believed that MIST1 does not initiate the transcription of these genes, but rather functions to enhance expression of genes required for regulated exocytosis. Therefore,

MIST1 has been described as a scaling factor. Importantly, reactivation of MIST1 in adult Mist1$^{-/-}$ tissue restores expression and acinar cell organization, suggesting this is a primary role for MIST1 function. The regulation of MIST1 is less understood. While it clearly is downstream of PTF1a and PDX1 in the pancreas, MIST1 is expressed in other serous exocrine tissues, including salivary glands and chief cells, and no studies have shown direct targeting of MIST1 by PTF1a or PDX1. However, both NR5A2 and X-box binding protein (XBP1) appear to directly regulate MIST1.

X-Box Binding Protein 1

X-box Binding Protein 1 (XBP1) is a bHLH transcription factor and a key mediator of the unfolded protein response (UPR) pathway, activated under conditions in which protein folding capacity of the cells is adversely affected. In non-stressed cells, XBP1 shows limited transcriptional ability. Upon activation of the UPR, Inositol-requiring enzyme 1 (IRE1) becomes active and deletes a 26 bp sequence from the Xbp1 mRNA, producing spliced Xbp1 transcript which encodes a transcription factor. sXbp1 targets genes involved in ER-associated degradation and increased protein folding as well as promoting apoptosis depending on the magnitude and length of the cell stress.

Likely due to the heavy protein folding capacity required in the pancreas, sXBP1 is present and functional even under physiological conditions. Targeted deletion of Xbp1 during development results in embryonic lethality. Liver specific rescue allows mice to survive for a short time postnatal to observe the phenotypic abnormalities that exist in other secretory organs. In the pancreas XBP1 deletion leads to incomplete development of acinar cells and as a result undergo apoptosis during development. Interestingly, XBP1 function is also required for maintenance of the acinar cell phenotype as induced deletion in adult cells leads to widespread apoptosis and regeneration specifically from acinar cells that escaped cre-mediated recombination. It appears that XBP1 targets several important factors required for acinar cell differentiation including Nr5a2 and Mist1.

Signalling Pathways Involved in Pancreatic Development

While many signaling pathways have been identified to regulate pancreatic development, we have focused on a few key pathways that have been identified in many studies.

B-catenin/Wnt

Wnt (Wingless-related integration site) signalling controls many embryonic development processes including body axis patterning, cell fate specification, and cell proliferation. Wnt signalling encompasess three different signaling pathways including the canonical or β-catenin-mediated pathway, and two non-canonical pathways that involve planar cell polarity and calcium. Signaling through the canonical Wnt pathway regulates expression of key mediators in pancreatic development, while the non-canonical Wnt pathways regulate cell shape and intracellular calcium regulation. In pancreatic development, several WNT ligands are expressed including WNT2b, WNT4, WNT5a and WNT7b. Binding of WNT ligands to Frizzled (Fzd) receptors leads to Fzd interaction with Dishevelled (Dvl), at which point the WNT signaling pathway diverges into one of the three pathways mentioned above. Activation of the canonical pathway leads to the dissociation of β-catenin from the APC/Axin/GSK3β complex, leading to β-catenin's translocation to the nucleus where it interacts with Lef-1/TCF (T-cell factor/lymphoid enhancer factor) transcription factors to affect gene expression.

The timing of β-catenin expression and activation of Wnt signalling is critical in pancreatic development. Typically, WNT signaling maintains a progenitor like state, thereby allowing expansion of the tissue prior to differentiation. Forced expression of WNT signalling prior to pancreatic specification prevents liver and pancreas development.

Detectable amounts of nuclear β-catenin are observed as early as E12.5-E14.5 in the pancreas, prior to the secondary transition. Activated β-catenin is critical during this time period as mice lacking β-catenin have smaller pancreata by E16.5. Targeted deletion of β-catenin results in the loss of exocrine pancreatic tissue along with increased tubular structures, increased parenchymal fibrosis and inflammatory infiltrate. The smaller pancreatic size in β-catenin null animals is due to a decrease in proliferation of PTF1A expressing progenitor cells and most likely not due to a loss of adherens junctions. These studies indicate that β-catenin is initially required for expansion of the pancreatic epithelium prior to differentiation, as well as for terminal differentiation of acinar cells. Although the exocrine pancreas was abolished, a loss of Wnt signalling did not disrupt the development of fully functional islets.

Retinoic Acid

Retinoic Acid (RA) is generated through the metabolism of Vitamin A by aldehyde dehydrogenases enzymes (ALDHs). RA is a ligand for the nuclear hormone receptor - retinoic acid receptor (RAR) - which exists as a DNA-bound repressor complex in the absence of RA. Heterodimers of RAR and Retinoid X Receptors (RXR) change conformation upon binding of RA to RAR, allowing recruitment of additional transcriptional regulators, thereby altering the expression of RAR-associated genes.

In zebrafish, blocking RA signalling results in undetectable expression levels of endocrine (insulin) or exocrine (trypsin) markers and specifies initiation of pancreatic progenitors upstream of Pdx1. In Xenopus, pharmacological and genetic approaches to block RA signaling diminished Pdx1 expression only in the dorsal pancreas, suggesting that RA signaling might be required only for endocrine pancreatic development in Xenopus. However, RA inhibition at later developmental stages blocked expression of differentiated exocrine markers but retained Pdx1. Therefore, in Xenopus, RA signaling is not required for specification of ventral pancreas but is necessary for later exocrine cell differentiation.

In mice, Raldh2 (Retinaldehyde dehydrogenase 2) is expressed between E8.75 and E12.5 in the dorsolateral mesenchyme that is in contact with the dorsal pancreatic bud. RA is synthesized in dorsolateral mesodermal cells, and the developing tissues responsiveness to RA follows a general dorsal to ventral gradient. Targeted deletion of Raldh2 in mice diminishes Pdx1 and Ptf1a expression in the presumptive dorsal pancreatic bud, while ventral pancreatic development is largely unaffected. Absence of Raldh2 during pancreatic development also decreases HB9 and ISL expression in the dorsal mesenchyme, supporting a mechanism by which dorsal bud agenesis may be due to deficient mesodermal signaling.

Hippo Signaling

Hippo (or Salvador/Warts/Hippo [SWH]) signaling regulated organ size through controlling cell proliferation and apoptosis. Hippo signaling involves activation of a Mst1/2 (Mammalian

Ste20-like kinases 1/2; Hippo in drosophila) and SAV1 (Save family WW domain-containing protein 1; Salvador in drosophila) complex, which leads to phosphorylation of LATS1/2 (Large tumour suppressor 1/2). LATS1/2 maintains YAP (Yes-associated protein) and TAZ (transcriptional co-activator with PDZ-binding motif) cytolasmic localization in an inactive conformation. Therefore, upon LATS1/2 phosphorylation, dephosphorylated YAP and TAZ translocate to the nucleus where they interact with TEAD1-4 and alter gene expression that promotes proliferation and inhibits apoptosis. Pancreata of Mst1/2 double knockout mice have a decreased pancreatic mass due to a failure in exocrine cell maturation. Additionally, overexpressing YAP in later pancreatic development expands the duct network while decreasing acinar and endocrine cells.

NOTCH Signaling

The NOTCH pathway plays a role in dictating different developmental fates for closely associated cells. Since duct, exocrine and endocrine cells arise from common epithelial compartments; NOTCH signaling has been suggested as a mechanism to ensure limited differentiation of endocrine cells, in neuronal development, NOTCH signaling triggers lateral inhibition, which prevents adjacent cells from having the same phenotype. In pancreatic development, cells expressing NGN3 are destined to become endocrine cells, and NGN3-expressing cells have enhanced expression of NOTCH ligands Delta (Dll), Jagged and Serrated. NOTCH signaling is triggered by interaction of these ligands with NOTCH receptors on adjacent cells resulting in receptor cleavage and release of the NOTCH intracellular domain (NICD). Translocation of NICD to the nucleus leads to modulation of transcriptional regulators mindbomb and RBPJ (recombination signal binding protein for immunoglobulin k J region), which enhances expression of Hairy Enhancer of Split (HES) 1, a repressor of Ngn3. Therefore, lateral inhibition through increased HES1 has been suggested as a way of preventing widespread endocrine differentiation. In support of this theory, deletion of Dll or Hes1 in the pancreas during development increased endocrine differentiation at the expense of the acinar cells. However, lateral inhibition cannot explain many of the effects of NOTCH signaling in pancreatic development, nor the final arrangement of exocrine vs. duct vs. endocrine cells. NOTCH signaling is time dependent, and forced early expression of NOTCH signaling also reduces the acinar cell population, implying a role for NOTCH in the expansion of this population of cells before exocrine differentiation.

Liver Development

Liver differentiation results from cellular interactions between the mesoderm and the endoderm. The presumptive hepatic endoderm is determined to differentiate into parenchymal cells from the 4-5 somite stage under the influence of the precardiac mesoderm. Later on, the endodermal cells proliferate and give rise to the hepatic buds from which strands of glandular cells are derived. The mesenchymal liver component stimulates both proliferation and differentiation of the endodermal cells. Its action is, however, not specific since the same effect is obtained with all the mesenchymes derived from the lateral plate mesoderm. The characteristics of the mesenchymal liver component were studied. In both quail and chick a cholinesterase activity was detected in the mesenchymal liver cells, which show an extensive development of the granular ergastoplasmic reticulum and of the Golgi apparatus. A close contact between endodermal and mesenchymal cells seems to be required for the cell interactions to occur. In monolayer cultures, glycogen synthesis ceases in the hepatocytes which after some days become fibroblastic in appearance. However, such a dedifferentiation is not irreversible. If reassociated with hepatic mesenchyme the fibroplastic cultures can recover their ability to synthesize and store glycogen.

Limb Development

Limb development in vertebrates is an area of active research in both developmental and evolutionary biology, with much of the latter work focused on the transition from fin to limb.

Limb formation begins in the morphogenetic limb field, as mesenchymal cells from the lateral plate mesoderm proliferate to the point that they cause the ectoderm above to bulge out, forming a limb bud. Fibroblast growth factor (FGF) induces the formation of an organizer at the end of the limb bud, called the apical ectodermal ridge (AER), which guides further development and controls cell death. Programmed cell death is necessary to eliminate webbing between digits.

The limb field is a region specified by expression of certain Hox genes, a subset of homeotic genes, and T-box transcription factors – Tbx5 for forelimb or wing development, and Tbx4 for leg or hindlimb development. Establishment of the forelimb field (but not hindlimb field) requires retinoic acid signaling in the developing trunk of the embryo from which the limb buds emerge. Also, although excess retinoic acid can alter limb patterning by ectopically activating Shh or Meis1/Meis2 expression, genetic studies in mouse that eliminate retinoic acid synthesis have shown that RA is not required for limb patterning.

The limb bud remains active throughout much of limb development as it stimulates the creation and positive feedback retention of two signaling regions: the AER and its subsequent creation of the zone of polarizing activity (ZPA) with the mesenchymal cells. In addition to the dorsal-ventral axis created by the ectodermal expression of competitive Wnt7a and BMP signals respectively, these AER and ZPA signaling centers are crucial to the proper formation of a limb that is correctly oriented with its corresponding axial polarity in the developing organism. Because these signaling systems reciprocally sustain each other's activity, limb development is essentially autonomous after these signaling regions have been established.

Limb Formation

Limb Bud

Limb formation begins in the morphogenetic limb field. Limb formation results from a series of epithelial-mesenchymal inductions between the mesenchymal cells of the lateral plate mesoderm and the overlying ectodermal cells. Cells from the lateral plate mesoderm and the myotome migrate to the limb field and proliferate to the point that they cause the ectoderm above to bulge out, forming the limb bud. The lateral plate cells produce the cartilaginous and skeletal portions of the limb while the myotome cells produce the muscle components.

The lateral plate mesodermal cells secrete fibroblast growth factors (FGF7 and FGF10) to induce the overlying ectoderm to form an organizer at the end of the limb bud, called the apical ectodermal ridge (AER), which guides further development and controls cell death. The AER secretes further growth factors FGF8 and FGF4 which maintain the FGF10 signal and induce proliferation in the mesoderm. The position of FGF10 expression is regulated by two Wnt signaling pathways: Wnt8c in the hindlimb and Wnt2b in the forelimb. The forelimb and the hindlimb are specified by their position along the anterior/posterior axis and possibly by two transcription factors: Tbx5 and Tbx4, respectively.

Precartilage Condensations

The limb's skeletal elements are prefigured by tight aggregates known as cellular condensations of the pre-cartilage mesenchymal cells. Mesenchymal condensation is mediated by extracellular matrix and cell adhesion molecules. In the process of chondrogenesis, chondrocytes differentiate from the condensations to form cartilage, giving rise to the skeletal primordia. In the development of most vertebrate limbs (though not in some amphibians), the cartilage skeleton is replaced by bone later in development.

Periodicities of the Limb Pattern

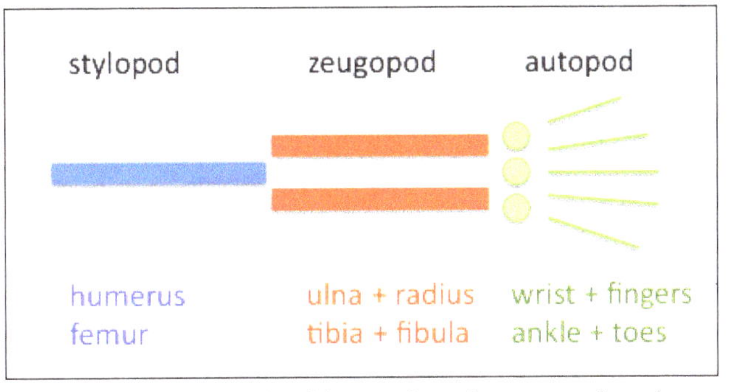

Vertebrate limbs are organized into stylopod, zeugopod, and autopod.

The limb is organized into three regions: stylopod, zeugopod, and autopod (in order from proximal to distal). The zeugopod and the autopod contain a number of periodic and quasi-periodic pattern motifs. The zeugopod consists of two parallel elements along the anteroposterior axis and the autopod contains 3-5 (in most cases) elements along the same axis. The digits also have a quasi-periodic arrangement along the proximodistal axis, consisting of tandem chains of skeletal elements. The generation of the basic limb plan during development results from the patterning of the mesenchyme by an interplay of factors that promote precartilage condensation and factors that inhibit it.

The development of the basic limb plan is accompanied by the generation of local differences between the elements. For example, the radius and ulna of the forelimb, and the tibia and fibula of the hindlimb of the zeugopod are distinct from one another, as are the different fingers or toes in the autopod. These differences can be treated schematically by considering how they are reflected in each of the limb's three main axes.

A general consensus is that the patterning of the limb skeleton involves one or more Turing-type reaction-diffusion mechanisms.

Evolution and Development

The evolution of limbs from paired fins has been an area of much focus. A reverse study of limb reduction and limb loss in the development of the snake is another active area of research. It has been shown that there are conserved sequences involved in limb development retained in the genome of the snake.

It is thought that these limb-enhancer sequences are conserved since there is overlap between those for limb development and those for phallus development (seen as a third type of appendage). This aspect has also been studied in the mouse where the usual limb-development signalling components are seen to play roles both in the development of the limbs and of the genital tubercle. The study of limb reduction and limb loss is unravelling the genetic pathways that control limb development.

The Turing system has enabled a number of possible outcomes in the evolutionary steps of patterning networks.

Axial Patterning and Related Issues

The developing limb has to align itself in relation to three axes of symmetry. These are the craniocaudal (head to tail), dorsoventral (back to front), and proximodistal (near to far) axes.

Many investigations into the development of the limb skeletal pattern have been influenced by the *positional information* concept proposed by Lewis Wolpert in 1971. In tune with this idea, efforts have been made to identify diffusive signaling molecules (morphogens) that traverse orthogonal axes of developing limbs and determine locations and identities of skeletal elements in a concentration-dependent fashion.

Proximodistal Patterning

Hox genes contribute to the specification of the stylopod, zeugopod and autopod. Mutations in Hox genes lead to proximodistal losses or abnormalities. Three different models have been advanced for explaining the patterning of these regions.

Progress Zone Model

The apical ectodermal ridge (AER) creates and maintains a zone of cell proliferation known as the progress zone. It is thought that cells here gain the positional information they need to travel to their destined position. It was proposed that their positional value was determined by the length of time that the cells were in the progress zone but this has yet to be proved. Proximal structures were proposed to be formed by the first cells to leave the zone and distal ones, by cells that left later.

The Progress Zone model was proposed 30 years ago but recent evidence has conflicted with this model.

Experimental evidence:

- Removing the AER at a later period of development results in less disruption of distal structures than if the AER was removed early in development.
- Grafting an early limb bud tip onto a late wing results in duplication of structures, while grafting a late wing bud tip onto an early limb results in a deletion of structures.

Early Allocation and Progenitor Expansion Model (or Prespecification Model)

Cells are specified for each segment in the early limb bud and this population of cells expand out as the limb bud grows. This model is consistent with the following observations. Cell division is seen throughout the limb bud. Cell death occurs within a 200 μm zone subjacent to the AER when it is removed; cell death forecloses some patterning. FGF-releasing beads are able to rescue limb development when the AER is removed by preventing this cell death.

Experimental evidence:

- Labeled cells in different position of an early limb bud were restricted to single segments of the limb.
- Limbs lacking expression of required FGF4 & FGF8 showed all structures of the limb and not just the proximal parts.

More recently, however, the investigators primarily responsible for both the Progress Zone and Prespecification models have acknowledged that neither of these models accounts adequately for the available experimental data.

Turing-type Reaction-diffusion Model

This model, a reaction–diffusion model first proposed in 1979, is based on the self-organizing properties of excitable media described by Alan Turing in 1952. The excitable medium is the limb bud mesenchyme, in which cells interact by positively autoregulatory morphogens such as transforming growth factor beta (TGF-β) and inhibitory signaling pathways involving fibroblast growth factor (FGF) and Notch signaling. Proximodistal and craniocaudal axes are not considered to be independently specified, but instead emerge by transitions in the number of parallel elements as the undifferentiated apical zone of the growing limb bud undergoes reshaping. This model only specifies a "bare bones" pattern. Other factors like sonic hedgehog (Shh) and Hox proteins, primary informational molecules in the other models, are proposed instead to play a fine-tuning role.

Experimental evidence:

- Limb mesenchymal cells, when dissociated and grown in culture or reintroduced within ectodermal "hulls" can recapitulate essential aspects of pattern formation, morphogenesis and differentiation.
- Peculiarities of the limb skeletal pattern in the mouse Doublefoot mutant are predicted outcomes of a Turing-type mechanism.
- Progressive reduction in distal Hox genes in a Gli3-null background results in progressively more severe polydactyly, displaying thinner and densely packed digits, suggesting (with the aid of computer modeling) that the dose of distal Hox genes modulates the period or wavelength of digits specified by a Turing-type mechanism.

Craniocaudal Patterning

The discovery, in 1957, of the zone of polarizing activity (ZPA) in the limb bud provided a model for understanding the patterning activity by the action of a morphogenic gradient of sonic hedgehog (Shh). Shh is recognised as a limb-specific enhancer. Shh is both sufficient and necessary to create the ZPA and specify the craniocaudal pattern in the distal limb (Shh is not necessary for the polarity of the stylopod). Shh is turned on in the posterior through the early expression of Hoxd genes, the expression of Hoxb8, and the expression dHAND. Shh is maintained in the posterior through a feedback loop between the ZPA and the AER. Shh induces the AER to produce FGF4 and FGF8 which maintains the expression of Shh.

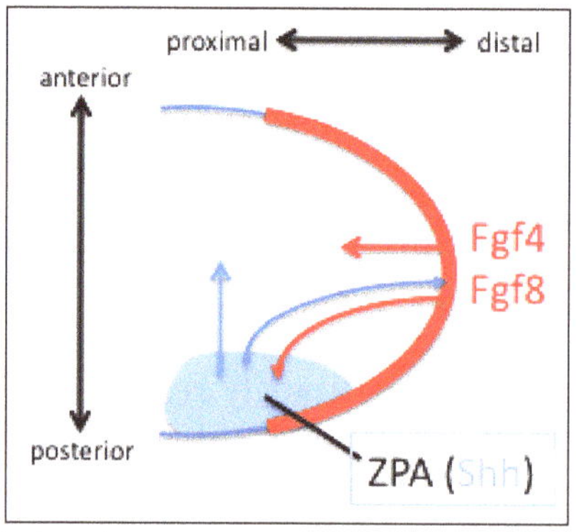

Early signals that define the craniocaudal (anterior-posterior), and proximodistal axes in vertebrate limb development.

Digits 3,4 and 5 are specified by a temporal gradient of Shh. Digit 2 is specified by a long-range diffusible form of Shh and Digit 1 does not require Shh. Shh cleaves the Ci/Gli3 transcriptional repressor complex to convert the transcription factor Gli3 to an activator which activates the transcription of HoxD genes along the craniocaudal. Loss of the Gli3 repressor leads to the formation of generic (non-individualized) digits in extra quantities.

Dorsoventral Patterning

Dorsoventral patterning arises from Wnt7a signals in the overlying ectoderm not the mesoderm. Wnt7a is both necessary and sufficient to dorsalize the limb. Wnt7a also influences the craniocaudal and loss of Wnt7a causes the dorsal side of limbs to become ventral sides and causes missing posterior digits. Replacing Wnt7a signals rescues this defect. Wnt7a is also required to maintain expression of Shh.

Wnt-7a also causes Lmx-1, a Lim Homebox gene (and thus a transcription factor), to be expressed. Lmx-1 is involved in dorsalisation of the limb, which was shown by knocking out the Lmx-1 gene in mice. The mice lacking the Lmx-1 produced ventral skin on both sides of their paws. There are other factors thought to control the DV patterning; Engrailed-1 represses the dorsalizing effect of Wnt-7a on the ventral side of the limbs.

Development of the Nervous System in Humans

The development of the nervous system in humans involves the studies of neuroscience and developmental biology or embryology to describe the cellular and molecular mechanisms by which complex nervous systems form and develop during prenatal development and continue to develop after childbirth.

Some landmarks of neural development in the embryo include the birth and differentiation of neurons from stem cell precursors (neurogenesis), the migration of immature neurons from their birthplaces in the embryo to their final positions, outgrowth of axons from neurons and guidance of the motile growth cone through the embryo towards postsynaptic partners, the generation of synapses between these axons and their postsynaptic partners, the synaptic pruning that occurs in adolescence, and finally the lifelong changes in synapses which are thought to underlie learning and memory.

Typically, these neurodevelopmental processes can be broadly divided into two classes: activity-independent mechanisms and activity-dependent mechanisms. Activity-independent mechanisms are generally believed to occur as hardwired processes determined by genetic programs played out within individual neurons. These include differentiation, migration and axon guidance to their initial target areas. These processes are thought of as being independent of neural activity and sensory experience. Once axons reach their target areas, activity-dependent mechanisms come into play. Neural activity and sensory experience will mediate formation of new synapses, as well as synaptic plasticity, which will be responsible for refinement of the nascent neural circuits.

Development of the Brain

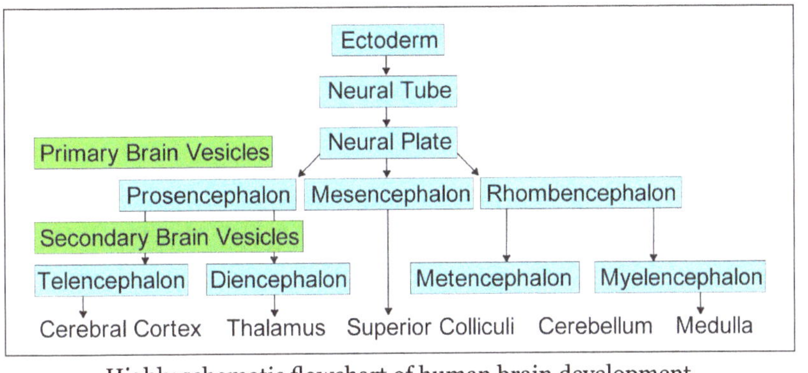

Highly schematic flowchart of human brain development.

Neuronal Migration

Neuronal migration is the method by which neurons travel from their origin or birthplace to their final position in the brain. Their most common means of migration are radial and tangential migration.

Radial Migration

Neural stem cells proliferate in the ventricular zone of the developing neocortex. The first postmitotic cells to migrate from the preplate which are destined to become Cajal-Retzius cells and subplate neurons. These cells do so by somal translocation. Neurons migrating with this mode of locomotion are bipolar and attach the leading edge of the process to the pia. The soma is then transported to the pial surface by nucleokenisis, a process by which a microtubule "cage" around the nucleus elongates and contracts in association with the centrosome to guide the nucleus to its final destination. Radial fibres (also known as radial glia) can translocate to the cortical plate and differentiate either into astrocytes or neurons. Somal translocation can occur at any time during development.

Subsequent waves of neurons split the preplate by migrating along radial glial fibres to form the cortical plate. Each wave of migrating cells travel past their predecessors forming layers in an inside-out manner, meaning that the youngest neurons are the closest to the surface. It is estimated that glial guided migration represents 80-90% of migrating neurons.

Axophilic Migration

Many neurons migrating along the anterior-posterior axis of the body use existing axon tracts to migrate along in a process called axophilic migration. An example of this mode of migration is in GnRH-expressing neurons, which make a long journey from their birthplace in the nose, through the forebrain, and into the hypothalamus. Many of the mechanisms of this migration have been worked out, starting with the extracellular guidance cues that trigger intracellular signaling. These intracellular signals, such as calcium signaling, lead to actin and microtubule cytoskeletal dynamics, which produce cellular forces that interact with the extracellular environment through cell adhesion proteins to cause the movement of these cells.

Tangential Migration

Most interneurons migrate tangentially through multiple modes of migration to reach their appropriate location in the cortex. An example of tangential migration is the movement of Cajal-Retzius cells within the marginal zone of the cortical neuroepithelium.

Others

There is also a method of neuronal migration called multipolar migration. This is seen in multipolar cells, which are abundantly present in the cortical intermediate zone. They do not resemble the cells migrating by locomotion or somal translocation. Instead these multipolar cells express neuronal markers and extend multiple thin processes in various directions independently of the radial glial fibers.

Neurotrophic Factors

Neurotrophic factors are molecules which promote and regulate neuronal survival in the developing nervous system. They are distinguished from ubiquitous metabolites necessary for cellular maintenance and growth by their specificity; each neurotrophic factor promotes the survival of only certain kinds of neurons during a particular stage of their development. In addition, it had been argued that neurotrophic factors are involved in many other aspects of neuronal development ranging from axonal guidance to regulation of neurotransmitter synthesis.

Adult Neurogenesis

Contrary to popular belief, neurogenesis also occurs in specific parts of the adult brain.

Adult Neural Development

Neurodevelopment in the adult nervous system includes mechanisms such as remyelination, generation of new neurons, glia, axons, myelin or synapses. Neuroregeneration differs between the peripheral nervous system (PNS) and the central nervous system (CNS) by the functional mechanisms and especially, the extent and speed.

Prenatal and Postnatal Development of the Human Nervous System

Almost all nerve cells, or neurons, are generated during prenatal life, and in most cases they are not replaced by new neurons thereafter. Morphologically, the nervous system first appears about 18 days after conception, with the genesis of a neural plate. Functionally, it appears with the first sign of a reflex activity during the second prenatal month, when stimulation by touch of the upper lip evokes a withdrawal response of the head. Many reflexes of the head, trunk, and extremities can be elicited in the third month.

During its development the nervous system undergoes remarkable changes to attain its complex organization. In order to produce the estimated 1 trillion neurons present in the mature brain, an average of 2.5 million neurons must be generated per minute during the entire prenatal life. This includes the formation of neuronal circuits comprising 100 trillion synapses, as each potential neuron is ultimately connected with either a selected set of other neurons or specific targets such as sensory endings. Moreover, synaptic connections with other neurons are made at precise locations on the cell membranes of target neurons. The totality of these events is not thought to be the exclusive product of the genetic code, for there are simply not enough genes to account for such complexity. Rather, the differentiation and subsequent development of embryonic cells into mature neurons and glial cells are achieved by two sets of influences:

(1) Specific subsets of genes and

(2) Environmental stimuli from within and outside the embryo.

Genetic influences are critical to the development of the nervous system in ordered and temporally timed sequences. Cell differentiation, for example, depends on a series of signals that regulate transcription, the process in which deoxyribonucleic acid (DNA) molecules give rise to ribonucleic acid (RNA) molecules, which in turn express the genetic messages that control cellular activity. Environmental influences derived from the embryo itself include cellular signals that consist of diffusible molecular factors. External environmental factors include nutrition, sensory experience, social interaction, and even learning. All of these are essential for the proper differentiation of individual neurons and for fine-tuning the details of synaptic connections. Thus, the nervous system requires continuous stimulation over an entire lifetime in order to sustain functional activity.

Neuronal Development

In the second week of prenatal life, the rapidly growing blastocyst (the bundle of cells into which a fertilized ovum divides) flattens into what is called the embryonic disk. The embryonic disk soon acquires three layers: the ectoderm (outer layer), mesoderm (middle layer), and endoderm (inner layer). Within the mesoderm grows the notochord, an axial rod that serves as a temporary backbone. Both the mesoderm and notochord release a chemical that instructs and induces adjacent undifferentiated ectoderm cells to thicken along what will become the dorsal midline of the body, forming the neural plate. The neural plate is composed of neural precursor cells, known as neuroepithelial cells, which develop into the neural tube. Neuroepithelial cells then commence to divide, diversify, and give rise to immature neurons and neuroglia, which in turn migrate from the neural tube to their final location. Each neuron forms dendrites and an axon; axons elongate and form branches, the terminals of which form synaptic connections with a select set of target neurons or muscle fibres.

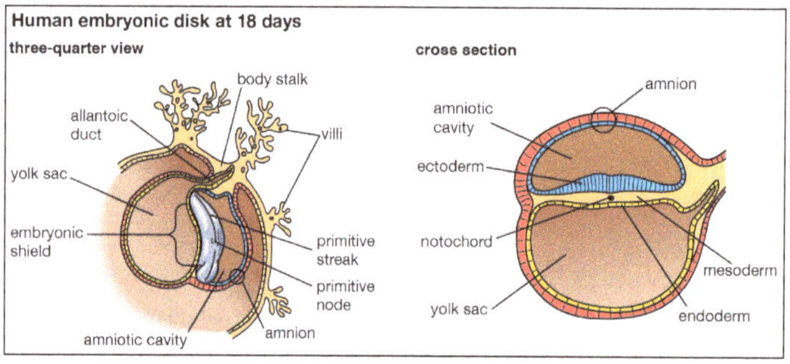

Development of the human embryo: Embryo of 18 days at disk or shield stage, (Ja) three-quarter view, and (Jb) cross section.

The remarkable events of this early development involve an orderly migration of billions of neurons, the growth of their axons (many of which extend widely throughout the brain), and the formation of thousands of synapses between individual axons and their target neurons. The migration and growth of neurons are dependent, at least in part, on chemical and physical influences. The growing tips of axons (called growth cones) apparently recognize and respond to various molecular signals, which guide axons and nerve branches to their appropriate targets and eliminate those that try to synapse with inappropriate targets. Once a synaptic connection has been established, a target cell releases a trophic factor (e.g., nerve growth factor) that is essential for the survival of the neuron synapsing with it. Physical guidance cues are involved in contact guidance, or the migration of immature neurons along a scaffold of glial fibres.

In some regions of the developing nervous system, synaptic contacts are not initially precise or stable and are followed later by an ordered reorganization, including the elimination of many cells and synapses. The instability of some synaptic connections persists until a so-called critical period is reached, prior to which environmental influences have a significant role in the proper differentiation of neurons and in fine-tuning many synaptic connections. Following the critical period, synaptic connections become stable and are unlikely to be altered by environmental influences. This suggests that certain skills and sensory activities can be influenced during development (including postnatal life), and for some intellectual skills this adaptability presumably persists into adulthood and late life.

Morphological Development

By 18 days after fertilization, the ectoderm of the embryonic disk thickens along what will become the dorsal midline of the body, forming the neural plate and, slightly later, the primordial eye, ear, and nose. The neural plate elongates, and its lateral edges rise and unite in the midline to form the neural tube, which will develop into the central nervous system. The neural tube detaches from the skin ectoderm and sinks beneath the surface. At this stage, groupings of ectodermal cells, called neural crests, develop as a column on each side of the neural tube. The cephalic (head) portion of the neural tube differentiates into the prosencephalon (forebrain), mesencephalon (midbrain), and rhombencephalon (hindbrain), and the caudal portion becomes the spinal cord. The neural crests develop into most of the elements (e.g., ganglia and nerves) of the peripheral nervous system. This stage is reached at the end of the first embryonic month.

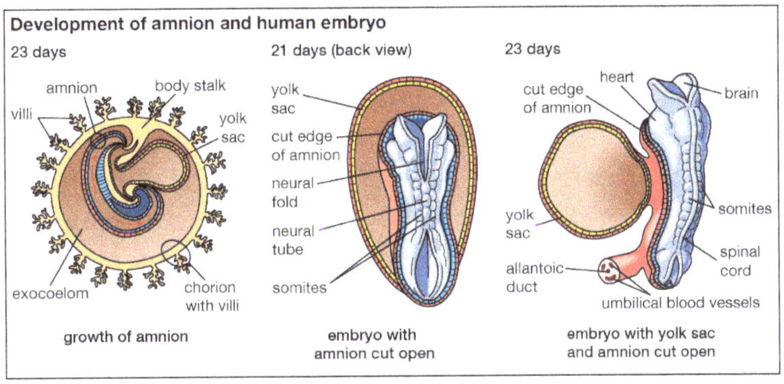

Development of the human embryo: Embryo of 23 days showing (K) growth of the amnion, (L) amnion cut open, and (M) yolk sac and amnion cut open.

The cells of the central nervous system originate from the ventricular zone of the neural tube—that is, the layer of neuroepithelial cells lining the central cavity of the tube. These cells differentiate and proliferate into neuroblasts, which are the precursors of neurons, and glioblasts, from which neuroglia develop. With a few exceptions, the neuroblasts, glioblasts, and their derived cells do not divide and multiply once they have migrated from the ventricular zone into the gray and white matter of the nervous system. Most neurons are generated before birth, although not all are fully differentiated. (One exception is the neurons of the olfactory nerve, which are generated continuously throughout life.) This effectively implies that an individual is born with a full complement of nerve cells.

By mid-fetal life the slender primordial brain of the neural-tube stage differentiates into a globular-shaped brain. Although fully mature size and shape are not obtained until puberty, the main outlines of the brain are recognizable by the end of the third fetal month. This early development is the product of several factors: the formation of three flexures (cephalic, pontine, and cervical); the differential enlargement of various regions, especially the cerebrum and the cerebellum; the massive growth of the cerebral hemispheres over the sides of the midbrain and of the cerebellum at the hindbrain; and the formations of convolutions (sulci and gyri) in the cerebral cortex and folia of the cerebellar cortex. The central and calcarine sulci are discernible by the fifth fetal month, and all major gyri and sulci are normally present by the seventh month. Many minor sulci and gyri appear after birth.

Postnatal Changes

The postnatal growth of the human brain is rapid and massive, especially during the first two years. By two years after birth, the size of the brain and the proportion of its parts are basically those of an adult. The typical brain of a full-term infant weighs 350 grams (12 ounces) at birth, 1,000 grams at the end of the first year, about 1,300 grams at puberty, and about 1,500 grams at adulthood. This increase is attributable mainly to the growth of preexisting neurons, new glial cells, and the myelination of axons. The trebling of weight during the first year (a growth rate unique to humans) may be an adaptation that is essential to the survival of humans as a species with a large brain. Birth occurs at a developmental stage when the infant is not so helpless as to be unable to survive, yet is small enough to be delivered out of the maternal pelvis. If the brain was much larger (enough, say, to support intelligent behaviour), normal delivery would not be possible.

Brain development in humans is thought to continue into the mid-20s, on average. From childhood into adulthood, processes such as synaptic pruning, the formation of new neural connections, and the strengthening of established connections shape brain development. Those same processes, which underlie neuroplasticity, may also influence brain changes later in adulthood. Nonetheless, in adulthood, as in early brain development, neurons that are not fired or used atrophy or die. In healthy adults, some 85,000 neurons in the neocortex may be lost each day. By age 75, the weight of the brain is reduced from its maximum at maturity by about one-tenth, the flow of blood through the brain by almost one-fifth, and the number of functional taste buds by about two-thirds. A loss of neurons does not necessarily imply a comparable loss of function; however, some loss may be compensated for by the formation from viable neurons of new branches of nerve fibres and by the formation of new synapses.

References

- A-Organogenesis-and-Vertebrate-Formation, Animal-Reproduction-and-Development, General-Biology, Introductory-and-General-Biology,: libretexts.org, Retrieved 15 August, 2019
- Edlund, Helena (July 2002). "Organogenesis: Pancreatic organogenesis — developmental mechanisms and implications for therapy". Nature Reviews Genetics. 3 (7): 524–532. doi:10.1038/nrg841. ISSN 1471-0064. PMID 12094230
- Science: sciencedirect.com, Retrieved 17 January, 2019
- Development-of-pancreas: pancreapedia.org, Retrieved 19 Feb February, 2019
- Schöler, Hans R. (2007). "The Potential of Stem Cells: An Inventory". In Nikolaus Knoepffler; Dagmar Schipanski; Stefan Lorenz Sorgner (eds.). Humanbiotechnology as Social Challenge. Ashgate Publishing. p. 28. ISBN 978-0-7546-5755-2
- Human-nervous-system, science: britannica.com, Retrieved 21 March, 2019
- Wolpert L (1971). Positional information and pattern formation. Curr Top Dev Biol. Current Topics in Developmental Biology. 6. pp. 183–224. doi:10.1016/S0070-2153(08)60641-9. ISBN 9780121531065. PMID 4950136
- Chen JP, Zhang DW, Liu JM (July 2016). "Exotic skyrmion crystals in chiral magnets with compass anisotropy". Scientific Reports. 6:29126. Bibcode:2016NatSR. 629126C. doi:10.1038/srep29126. PMC 4932608. PMID 27377149

4

Plant Development

The process by which structures originate and mature as a plant grows is known as plant development. Plants continuously produce new structures and tissues from a tissue known as meristem. The topics elaborated in this chapter will help in gaining a better perspective about the processes related to plant growth and development as well as its stages.

Plant development is a multiphasic process in which two distinct plant forms succeed each other in alternating generations. One form, the sporophyte, is created by the union of gametes (sex cells) and is thus diploid (contains two sets of similar chromosomes). At maturity, the sporophyte produces haploid (containing a single set of chromosomes) spores, which grow into the gametophyte generation. At their sexual maturity, the gametophytes produce haploid gametes that unite to begin a new cycle.

Although both plants and animals share the chemical basis of inheritance and of translation of the genetic code into structural units called proteins, plant development differs from that of animals in several important ways. Higher plants sustain growth throughout life and, in this sense, are perpetually embryonic; animals, on the other hand, generally have a determinate period of growth, after which they are considered mature. Furthermore, both growth and organ formation in plants are influenced by their possession of a rigid cell wall and a fluid-filled space called the vacuole, two features unique to the plant cell. Conversely, certain features of animal cells are absent in plants. Notable is the lack of cellular movements and fusions that play an important part in tissue and organ development in higher animals.

General Features

Life Cycles

The life cycle of all tracheophytes (vascular plants), bryophytes (mosses and liverworts), and many algae and fungi are based on an alternation of generations, or different life phases: the gametophyte, which produces gametes, or sex cells, alternating with the sporophyte, which produces spores. Gametophytes develop from the spores and, like them, are normally haploid; i.e., each cell has one set of chromosomes. Sporophytes develop from a fertilized egg, or zygote that results from the fusion of gametes (fertilization) formed by the gametophytes and is accordingly diploid; i.e., each cell has two sets of chromosomes. Although the two generations are phases of one life cycle, they have independent developmental histories; each begins as a single cell, passes through a juvenile period, matures, and gives rise to the alternate phase.

The alternating generations typically have different forms (i.e., are heteromorphic); this is true for the bryophytes and for all vascular plants, including lower vascular plants (ferns and allies), angiosperms (flowering plants), and gymnosperms (conifers and allies). General rules for vascular plants are that the sporophyte generation is physically the larger, has a more complex developmental history, produces a greater range of cell types, and expresses a more diverse biochemistry; the gametophyte is often diminutive, reduced in the case of the angiosperms to a mere few cells. In the bryophytes, the gametophyte generation, rather than the sporophyte, is the more conspicuous.

Gametophyte: A liverwort gametophyte

Although the gametophyte generation in vascular plants is small and has limited physiological capabilities, its cells must convey genes capable of directing the sporophytic developmental pattern, because the pattern is transmitted through the gametes to the zygote. The expression of "sporophytic" genes must therefore be repressed in the gametophyte, probably from the time of spore formation (sporogenesis). Correspondingly, events associated with gamete formation (gametogenesis) or fertilization must somehow free the sporophytic genes and thus permit the zygote to enter the sporophytic developmental pattern. Although it might be supposed that the "switch" is associated with the difference in chromosome number between the haploid spore (a single set) and the diploid zygote (a double set), this has been shown not to be the determining factor.

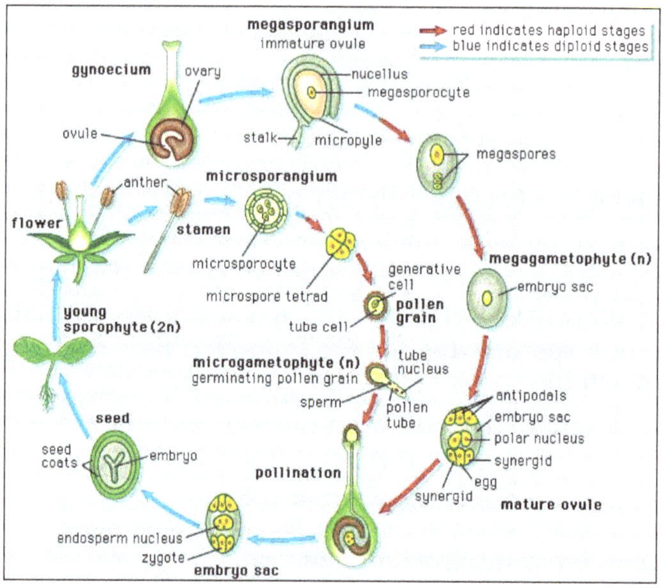

Life cycle of a typical angiosperm

In figure, the angiosperm life cycle consists of a sporophyte phase and a gametophyte phase. The cells of a sporophyte body have a full complement of chromosomes (i.e., the cells are diploid, or

2n); the sporophyte is the typical plant body that one sees when one looks at an angiosperm. The gametophyte arises when cells of the sporophyte, in preparation for reproduction, undergo meiotic division and produce reproductive cells that have only half the number of chromosomes (i.e., haploid, or n). A two-celled microgametophyte (called a pollen grain) germinates into a pollen tube and through division produces the haploid sperm. An eight-celled megagametophyte (called the embryo sac) produces the egg. Fertilization occurs with the fusion of a sperm with an egg to produce a zygote, which eventually develops into an embryo. After fertilization, the ovule develops into a seed, and the ovary develops into a fruit.

The alternation of generations illustrates an important principle, namely that cell lineages arising from single parental cells containing the same genetic potentiality may pursue mutually exclusive developmental patterns. Channeling, or canalizing, events of this nature occur repeatedly in the course of development of an individual plant, beginning with the pattern of cell division from the very first cleavage of the zygote cell.

Body Plans

Collectively, plants manifest a wide range of body plans, ranging from small multicellular structures to enormous trees. Among nonvascular plants, true parenchyma is found in the bryophytes, in both the gametophyte and sporophyte phases. The development of the moss gametophyte illustrates the transition from a filamentous to a highly organized three-dimensional growth form. The moss spore germinates into a filamentous plant, the protonema, which later produces a leafy shoot. This type of transition from simple to more complex growth form is accompanied by the synthesis of new kinds of ribonucleic acids (RNA's), presumably through the activation of genes that were not expressed during the early growth of the gametophyte.

The development of the complex body forms of vascular-plant sporophytes, which do not normally pass through any filamentous stages. It may be noted, however, that, in the course of evolution, the capacity for this type of growth has not been lost, since it may be adopted by cells grown in tissue cultures in the laboratory.

Preparatory Events

The sporophytes of all vascular plants produce cells called spore mother cells—since they will give rise to spores—in spore cases (sporangia). Spore mother cells are usually surrounded, during development, by a special nutritive tissue. In the more primitive groups, each sporangium holds many mother cells. This is true also in the pollen-producing sporangia of gymnosperms and angiosperms but not in the egg-producing sporangia (ovules), which usually have only one mother cell.

In certain lower vascular plants, typified by the spike moss Selaginella, the gametophyte is formed entirely—or almost entirely—within the spore wall. Two kinds of gametophytes develop from the two kinds of spores produced by the sporophyte in different sporangia; the larger spore (megaspore) gives rise to the female gametophyte, the smaller spore (microspore) to the male. This condition is referred to as heterospory. The gametophytes, or prothalli, of other club mosses and most horsetails and ferns are sexually undifferentiated and arise from one kind of spore, a condition termed homospory.

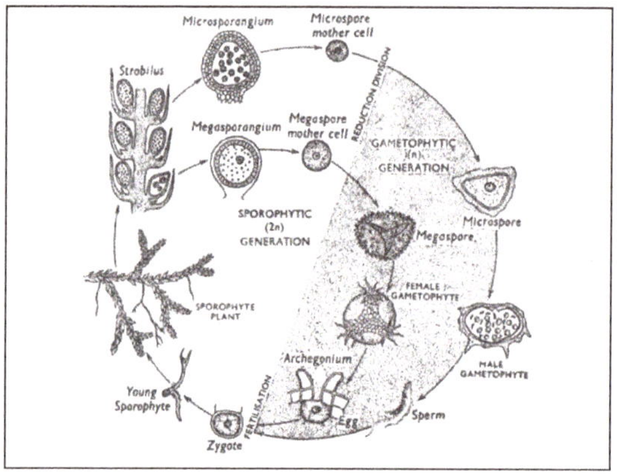

Life cycle of Selaginella: The only genus of spike moss.

In these homosporous groups, the gametophytes develop as free-living and independent plants that ultimately produce the gametes. In general, the male gametes (antherozoids) are produced in globose structures (antheridia) that are either stalked or sunken in the gametophyte. The antherozoids, always many in number, develop from mother cells enclosed in the jacket of the antheridium. Each antherozoid can move by using its whiplike hairs, or flagella, two or three (in the lycopods) or many (in the horsetails and ferns). The female gametes are formed singly in flask-shaped structures (archegonia) that also are either stalked or sunken in the gametophyte. The neck of the flask is closed by neck canal cells, which later break down to permit the entry of the male gamete. The egg itself lies in the basal part, or venter, of the flask, with a ventral canal cell above it. When the male gametes, or antherozoids, are released by the rupture of the antheridium, they swim in a water film to the archegonia and effect fertilization.

Among the gymnosperms the male gametophyte is much reduced and is a parasite on the sporophyte for only a short time. Cell cleavages within the spore wall cut off a prothallial cell, which will give rise to the vegetative (i.e., nonreproductive) part of the plant, and an antheridial cell, which divides into a tube cell and a generative cell. The male gametophyte so formed and contained within the spore wall is the pollen grain. After transfer to the ovule by wind, the pollen grain germinates to form a tube, and the generative cell divides into two cells, one of which forms the male gametes by further division. The gametes bear numerous spirally arranged flagella. The female gametophyte meanwhile develops entirely within the parent sporangium in the ovule. The size of the single functional spore increases greatly as the spore nucleus divides repeatedly to produce numerous free nuclei. Cell-wall formation then begins at the periphery, extending inward until the whole area is divided into cells. Up to four archegonia are formed, sunken in the tissue of the gametophyte, each with a female gamete, or egg.

Wind pollination: Windblown pollen from the male cone of a lodgepole pine (Pinus contorta).

The end of the gametophyte phase and the beginning of the sporophyte phase occur at fertilization, when one of the male gametes fuses with the female gamete to form the zygote, which will then develop as the sporophyte. (Development of the sporophyte can, in some cases, be triggered by means other than fertilization, in which case the organism is said to arise parthenogenetically).

The male gametophyte of angiosperms is reduced to three cells, one so-called vegetative cell and two male gametes. The division producing the gametes may occur either before dispersal of the pollen grain or later, during the growth of the pollen tube. The female sporangium has one or two coats, or integuments, except for an opening (micropyle) at one end; the sporangium with an integument is called the ovule. The female gametophyte, known in this group as the embryo sac, develops from the parent spore while it is still retained in the sporangium. Three cell divisions result in eight nuclei, which arrange themselves so that three lie at each end and two lie in the centre. The cytoplasm then cleaves and three cells are formed at each pole, leaving two nuclei in a large central cell. The three cells at the micropylar pole (end toward the micropyle) form the egg apparatus. Two of these cells, called synergids, correspond to the neck cells of an archegonium; the third is the egg cell. The three cells at the opposite pole, the antipodals, play a part in embryo nutrition in certain genera. The two polar nuclei in the central cell ultimately unite, becoming the fusion nucleus. The pollen grain is transferred by various agencies (wind, water, animals) to the stigma of the female flower, and, as in the gymnosperms, it germinates to produce a tube. This tube grows through intervening tissues, through an opening (micropyle) of the egg, and enters a cell near the micropyle (synergid), in which the two male gametes are discharged. The unique feature of this phase of angiosperm development is that two fertilizations occur. One male gamete fuses with the egg to give the diploid zygote; the other makes its way to the fusion nucleus in the central cell, already diploid, and by a second fusion gives a triploid primary endosperm nucleus, which is later concerned in the formation of the nutritive tissue, or endosperm.

Plants and animals are separated by about 1.5 billion years of evolutionary history. They have evolved their multicellular organization independently but using the same initial tool kit—the set of genes inherited from their common unicellular eucaryotic ancestor. Most of the contrasts in their developmental strategies spring from two basic peculiarities of plants. First, they get their energy from sunlight, not by ingesting other organisms. This dictates a body plan different from that of animals. Second, their cells are encased in semirigid cell walls and cemented together, preventing them from moving as animal cells do. This dictates a different set of mechanisms for shaping the body and different developmental processes to cope with a changeable environment.

Animal development is largely buffered against environmental changes, and the embryo generates the same genetically determined body structure unaffected by external conditions. The development of most plants, by contrast, is dramatically influenced by the environment. Because they cannot match themselves to their environment by moving from place to place, plants adapt instead by altering the course of their development. Their strategy is opportunistic. A given type of organ—a leaf, a flower, or a root, say—can be produced from the fertilized egg by many different paths according to environmental cues. A begonia leaf pegged to the ground may sprout a root; the root may throw up a shoot; the shoot, given sunlight, may grow leaves and flowers.

The mature plant is typically made of many copies of a small set of standardized modules, as described in figure. The positions and times at which those modules are generated are strongly influenced by the environment, causing the overall structure of the plant to vary. The choices between alternative modules and their organization into a whole plant depend on external cues and long-range hormonal signals that play a much smaller part in the control of animal development.

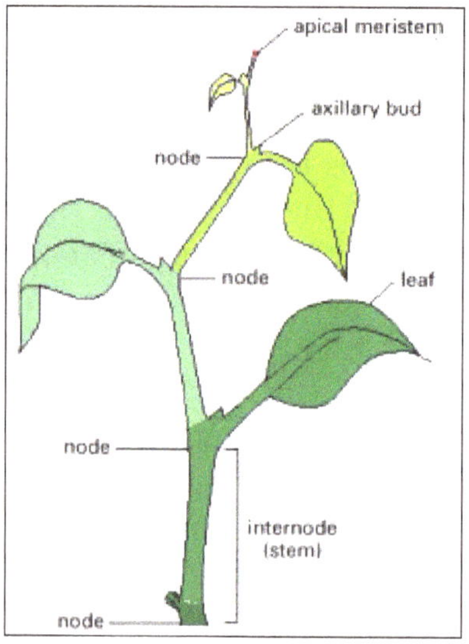

In figure showing, simple example of the modular construction of plants. Each module (shown in different shades of green) consists of a stem, a leaf, and a bud containing a potential growth center, or meristem. The bud forms at the branch point, or node, where the leaf diverges from the stem. Modules arise sequentially from the continuous activity of the apical meristem.

But although the global structure of a plant—its pattern of roots or branches, its numbers of leaves or flowers—can be highly variable, its detailed organization on a small scale is not. A leaf, a flower, or indeed an early plant embryo, is as precisely specified as any organ of an animal, possessing a determinate structure, in contrast with the indeterminate pattern of branching and sprouting of the plant as a whole. The internal organization of a plant module raises essentially the same problems in the genetic control of pattern formation as does animal development, and they are solved in analogous ways.

Arabidopsis Serves as a Model Organism for Plant Molecular Genetics

Flowering plants, despite their amazing variety, are of relatively recent origin. The earliest known fossil examples are 125 million years old, as against 350 million years or more for vertebrate animals. Underlying the diversity of form, therefore, there is a high degree of similarity in molecular mechanisms. As we shall see, a small genetic change can transform a plant's large-scale structure; and just as plant physiology allows survival in many different environments, so also it allows survival of many differently structured forms. A mutation that gives an animal two heads is generally lethal; one that doubles the number of flowers or branches on a plant is generally not.

To identify the genes that govern plant development and to discover how they function, plant biologists have selected a small weed, the common wall cress Arabidopsis thaliana (Figure), as their primary model organism. Like Drosophila or Caenorhabditis elegans, it is small, quick to reproduce, and convenient for genetics. It can be grown indoors in Petri dishes or tiny plant pots in large numbers and produces hundreds of seeds per plant after 8 to 10 weeks. It has, in common with C. elegans, a significant advantage over Drosophila or vertebrate animals for genetics: like many flowering plants, it can reproduce as a hermaphrodite because a single flower produces both eggs and the male gametes that can fertilize them. Therefore, when a flower that is heterozygous for a recessive lethal mutation is self-fertilized, one-fourth of its seeds will display the homozygous embryonic phenotype. This makes it easy to perform genetic screens and so to obtain a catalog of the genes required for specific developmental processes.

Arabidopsis thaliana: This small plant is a member of the mustard (or crucifer) family. It is a weed of no economic use but of great value for genetic studies of plant development.

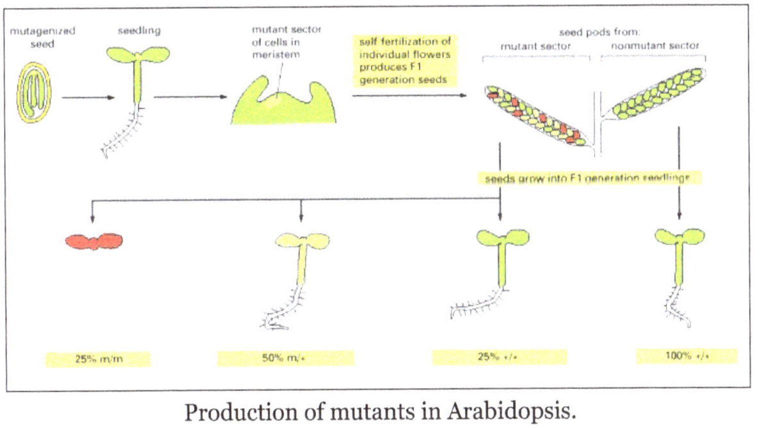

Production of mutants in Arabidopsis.

In figure, a seed, containing a multicellular embryo, is treated with a chemical mutagen and left to grow into a plant. In general, this plant will be a mosaic of clones of cells carrying different induced mutations. An individual flower produced by this plant will usually be composed of cells belonging to the same clone, all carrying the same mutation, m, in heterozygous form (m/+). Self-fertilization of individual flowers by their own pollen results in seed pods, each of which contains a family of embryos of whose members half, on average, will be heterozygous (m/+), one quarter will be homozygous mutant (m/m), and one quarter will be homozygous wild-type (+/+). Often, the mutation will have a recessive lethal effect, as indicated here by the lack of a root in the m/m seedling. The mutant stock is then maintained by breeding from the heterozygote: it will produce seed pods (F2 generation) that all contain a mixture of +/+, m/+, and m/m seeds.

The Arabidopsis Genome is Rich in Developmental Control Genes

Arabidopsis has one of the smallest plant genomes—125 million nucleotide pairs, on a par with C. elegans and Drosophila—and the complete DNA sequence is now known. It contains approximately 26,000 genes. This total includes many recently generated duplicates, however, so that the number of functionally distinct types of protein represented may be considerably less. Cell culture and genetic transformation methods have been established, as well as vast libraries of seeds carrying mutations produced by random insertions of mobile genetic elements, so that plants with mutations in any chosen gene can be obtained to order. Powerful tools are thus available to analyze gene functions. Although only a small fraction of the total gene set has been characterized experimentally as yet, functions can be tentatively assigned to many genes—about 18,000—on the basis of their sequence similarities to well-characterized genes in Arabidopsis and other organisms.

Even more than the genomes of multicellular animals, the Arabidopsis genome is rich in genes that code for gene regulatory proteins. Some major families of animal gene regulatory proteins (such as the Myb family of DNA-binding proteins) are greatly expanded, while others (such as nuclear hormone receptors) seem to be entirely absent, and there are large families of gene regulatory proteins in the plant that have no animal homologs.

Table: Some Major Families of Gene Regulatory Proteins in Arabidopsis, Drosophila, C. elegans, and the Yeast Saccharomyces cerevisiae.

Family	Number of family members predicted from genome analysis			
	Arabidopsis	Drosophila	C. Elegans	Yeast
Myb	190	6	3	10
AP2/EREBP (Apetala2/ethylene-responsive-element binding protein)	144	0	0	0
bHLH (basic helix-loop-helix)	139	46	25	8
NAC	109	0	0	0
C2H2 (Zn finger)	105	291	139	53
Homeobox	89	103	84	9
MADS box	82	2	2	4
bZIP	81	21	25	21
WRKY (Zn finger)	72	0	0	0
GARP	56	0	0	0
C2C2 (Zn finger)/GATA	104	6	9	10
Nuclear hormone receptor	0	21	25	0
C6 (Zn finger)	0	0	0	52
Estimated total (including many not listed above)	1,533	635	669	209
% of genes in genome	5.9	4.5	3.5	3.5

Where homologous gene regulatory proteins (such as homeodomain proteins) can be recognized in both plants and animals, they have little in common with regard to the genes they regulate or the types of developmental decisions that they control, and there is very little conservation of protein sequence outside the DNA-binding domains.

Arabidopsis is like multicellular animals in possessing many genes for cell communication and signal transduction (1,900 genes out of 18,000 classified), but the specific details of these gene sets are very different, as The Wnt, Hedgehog, Notch, and TGFβ signaling mechanisms are all absent in Arabidopsis. In compensation, other signaling pathways peculiar to plants are highly developed. Cell-surface receptors of the tyrosine kinase class seem to be entirely absent, although many of the signaling components downstream of these receptors in animals are present. Conversely, receptors of the serine/threonine kinase class are very plentiful, but they do not act through the same system of intracellular messengers as the receptor serine/threonine kinases in animals. Substantial sets of genes are devoted to developmental processes of special importance in plants: more than 700 for synthesis and remodelling of the plant cell wall, for example, and more than 100 for detecting and responding to light.

1. We must now examine how the genes of the plant are used to control plant development.

2. Embryonic development starts by establishing a root-shoot axis and then halts inside the seed.

The basic strategy of sexual reproduction in flowering plants is briefly summarized in figure. The fertilized egg, or zygote, of a higher plant begins by dividing asymmetrically to establish the polarity of the future embryo. One product of this division is a small cell with dense cytoplasm, which will become the embryo proper. The other is a large vacuolated cell that divides further and forms a structure called the suspensor, which in some ways is comparable to the umbilical cord in mammals. The suspensor attaches the embryo to the adjacent nutritive tissue and provides a pathway for the transport of nutrients.

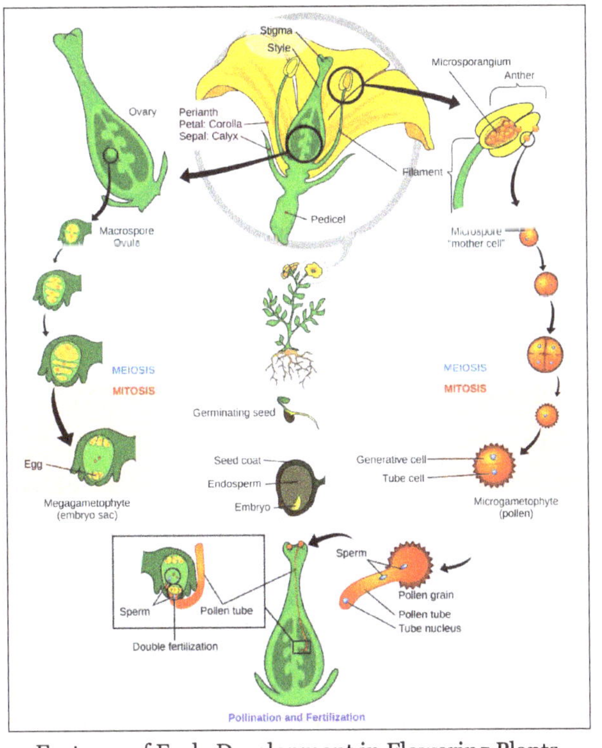

Features of Early Development in Flowering Plants.

During the next step in development the diploid embryo cell proliferates to form a ball of cells that quickly acquires a polarized structure. This comprises two key groups of proliferating cells—one at the suspensor end of the embryo that will collaborate with the uppermost suspensor cell to generate a root, and one at the opposite end that will generate a shoot. The main root-shoot axis established in this way is analogous to the head-to-tail axis of an animal. At the same time it begins to be possible to distinguish the future epidermal cells, forming the outermost layer of the embryo, the future ground tissue cells, occupying most of the interior, and the future vascular tissue cells, forming the central core. These three sets of cells can be compared to the three germ layers of an animal embryo. Slightly later in development, the rudiment of the shoot begins to produce the embryonic seed leaves, or cotyledons—one in the case of monocots and two in the case of dicots. Soon after this stage, development usually halts and the embryo becomes packaged in a seed (a case formed by tissues of the mother plant), specialized for dispersal and for survival in harsh conditions. The embryo in a seed is stabilized by dehydration, and it can remain dormant for a very long time—even hundreds of years. When rehydrated, the seeds germinate and embryonic development resumes.

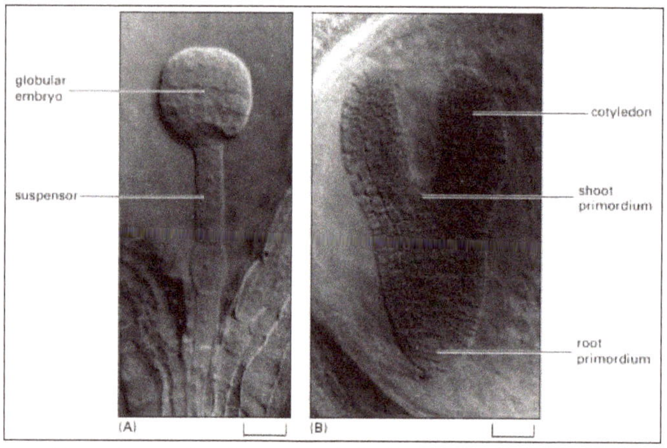

Two stages of embryogenesis in Arabidopsis thaliana.

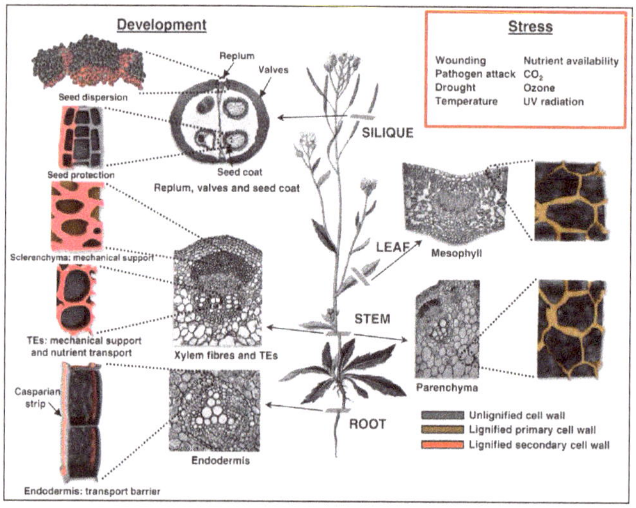

The cell types and tissues from which higher plants are constructed.

Genetic screens can be used in Arabidopsis, just as in Drosophila or C. elegans, to identify the genes that govern the organization of the embryo and to group these into categories according to their homozygous mutant phenotypes. Some are required for formation of the seedling root, some for the seedling stem, and some for the seedling apex with its cotyledons. Another class is required for formation of the three major tissue types—epidermis, ground tissue, and vascular tissue—and yet another class for the organized changes of cell shape that give the embryo and seedling their elongated form.

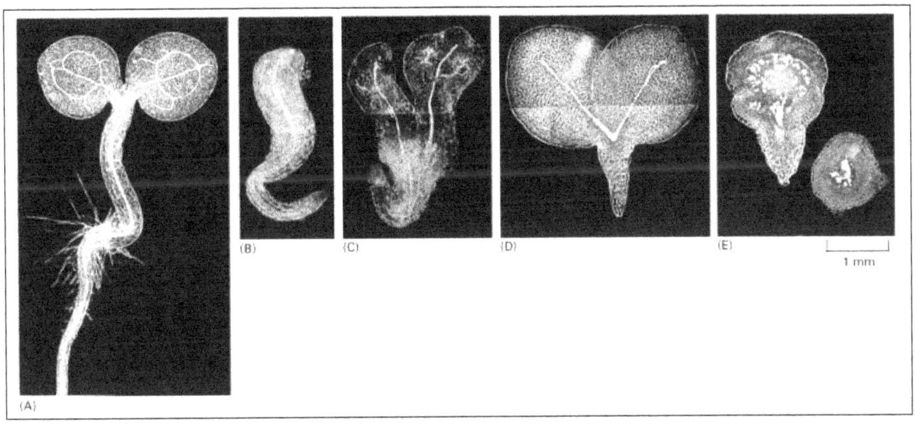

Mutant Arabidopsis seedlings.

In figure, A normal seedling (A) compared with four types of mutant (B-E) defective in different parts of their apico-basal pattern: (B) has structures missing at its apex, (C) has an apex and a root but lacks a stem between them, (D) lacks a root, and (E) forms stem tissues but is defective at both ends. The seedlings have been "cleared" so as to show the vascular tissue inside them (pale strands).

The Parts of a Plant are Generated Sequentially by Meristems

Roughly speaking, the embryo of an insect or a vertebrate animal is a rudimentary miniature scale model of the later organism, and the details of body structure are filled in progressively as it enlarges. The plant embryo grows into an adult in a quite different way: the parts of the adult plant are created sequentially by groups of cells that proliferate to lay down additional structures at the plant's periphery. These all-important groups of cells are called apical meristems. Each meristem consists of a self-renewing population of stem cells. As these divide, they leave behind a trail of progeny that become displaced from the meristem region, enlarge, and finally differentiate. Although the shoot and root apical meristems generate all the basic varieties of cells that are needed to build leaves, roots, and stems, many cells outside the apical meristems also keep a capacity for further proliferation and retain meristem potential. In this way trees and other perennial plants, for example, are able to increase the girth of their stems and roots as the years go by and can sprout new shoots from dormant regions if the plant is damaged.

The rudiments of the apical meristems of root and shoot are already determined in the embryo. As soon as the seed coat ruptures during germination, a dramatic enlargement of non-meristematic cells occurs, driving the emergence first of a root, to establish an immediate foothold in the soil, and then of a shoot. This is accompanied by rapid and continual cell divisions in the apical meristems: in the apical meristem of a maize root, for example, cells divide every 12 hours, producing 5 × 105 cells per day. The rapidly growing root and shoot probe the environment—the root increasing the plant's capacity for taking up water and minerals from the soil, the shoot increasing its capacity for photosynthesis.

Seedling of Arabidopsis. The brown objects to the right of the young seedling are the two halves of the discarded seed coat.

Development of the Seedling Depends on Environmental Signals

From germination onward, the course of plant development is powerfully influenced by signals from the environment. The shoot has to push its way rapidly up through the soil, and must open its cotyledons and begin photosynthesis only after it has reached the light. The timing of this transition from rapid subterranean sprouting to illuminated growth cannot be genetically programmed, because the depth at which the seed is buried is unpredictable. The developmental switch is controlled instead by light, which, among other effects, acts on the seedling by inhibiting production of a class of plant hormones called *brassinosteroids*. Mutations in genes required for production or reception of the brassinosteroid signal cause the stem of the seedling to go green, slow its elongation, and open its cotyledons prematurely, while it is still in the dark.

The shaping of each new structure depends on oriented cell division and expansion.

Plant cells, imprisoned within their cell walls, cannot crawl about and cannot be shuffled as the plant grows; but they can divide, and they can swell, stretch, and bend. The morphogenesis of a developing plant therefore depends on orderly cell divisions followed by strictly oriented cell expansions. Most cells produced in the root-tip meristem, for example, go through three distinct phases of development—division, growth (elongation), and differentiation. These three steps, which overlap in both space and time, give rise to the characteristic architecture of a root tip. Although the process of cell differentiation often begins while a cell is still enlarging, it is comparatively easy to distinguish in a root tip a zone of cell division, a zone of oriented cell elongation (which accounts for the growth in length of the root), and a zone of cell differentiation.

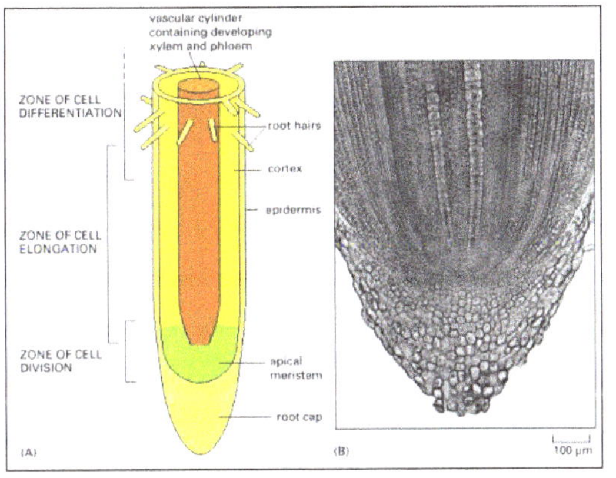

Growing root tip. (A) The organization of the final 2 mm of a growing root tip. The approximate zones in which cells can be found dividing, elongating, and differentiating are indicated. (B) The apical meristem and root cap of a corn root tip, showing the orderly files of cells produced.

In the phase of controlled expansion that generally follows cell division, the daughter cells may often increase in volume by a factor of 50 or more. This expansion is driven by an osmotically based turgor pressure that presses outward on the plant cell wall, and its direction is determined by the orientation of the cellulose fibrils in the cell wall, which constrain expansion along one axis. The orientation of the cellulose in turn is apparently controlled by the orientation of arrays of microtubules just inside the plasma membrane, which are thought to guide cellulose deposition. These orientations can be rapidly changed by plant growth regulators, such as ethylene and gibberellic acid, but the molecular mechanisms underlying these dramatic cytoskeletal rearrangements are still unknown.

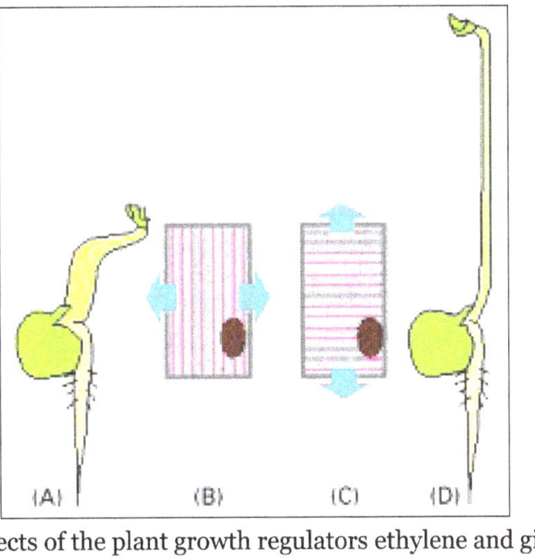

The different effects of the plant growth regulators ethylene and gibberellic acid.

In figure, these regulators exert rapid and opposing effects on the orientation of the cortical microtubule array in cells of young pea shoots. A typical cell in an ethylene-treated plant (B) shows a net longitudinal orientation of microtubules, while a typical cell in a gibberellic-acid-treated plant (C) shows a net transverse orientation. New cellulose microfibrils are deposited parallel to the microtubules. Since this influences the direction of cell expansion, gibberellic acid and ethylene encourage growth in opposing directions: ethylene-treated seedlings will develop short, fat shoots (A), while gibberellicacid-treated seedlings will develop long, thin shoots (D).

Each Plant Module Grows from a Microscopic Set of Primordia in a Meristem

The apical meristems are self-perpetuating: in a perennial plant, they carry on with their functions indefinitely, as long as the plant survives, and they are responsible for its continuous growth and development. But apical meristems also give rise to a second type of outgrowth, whose development is strictly limited and culminates in the formation of a structure such as a leaf or a flower, with a determinate size and shape and a short lifespan. Thus, as a vegetative (non-flowering) shoot elongates, its apical meristem lays down behind itself an orderly sequence of nodes, where leaves have grown out, and internodes (segments of stem). In this way the continuous activity of the meristem produces an ever increasing number of similar modules, each consisting of a stem, a leaf, and a bud. The modules are connected to one another by supportive and transport tissue, and successive modules are precisely located relative to each other, giving rise to a repetitively patterned structure. This iterative mode of development is characteristic of plants and is seen in many other structures besides the stem-leaf system.

Repetitive patterning in plants. Accurate placing of successive modules from a single apical meristem produces these elaborate but regular patterns in leaves (A), flowers (B), and fruits (C).

Although the final module is large, its organization, like that of an animal embryo, is mapped out at first on a microscopic scale. At the apex of the shoot, within a space of a millimeter or less, one finds a small, low central dome surrounded by a set of distinctive swellings in various stages of enlargement. The central dome is the apical meristem itself; each of the surrounding swellings is the primordium of a leaf. This small region, therefore, contains the already distinct rudiments of several entire modules. Through a well-defined program of cell proliferation and cell enlargement, each leaf primordium and its adjacent cells will grow to form a leaf, a node, and an internode. Meanwhile, the apical meristem itself will give rise to new leaf primordia, so as to generate more and more modules in a potentially unending succession. The serial organization of the modules of the plant is thus controlled by events at the shoot apex. It is the system of local signals within this tiny region that determines the pattern of primordia—the position of one leaf rudiment relative to the next, the spacing between them, and their location relative to the apical meristem itself.

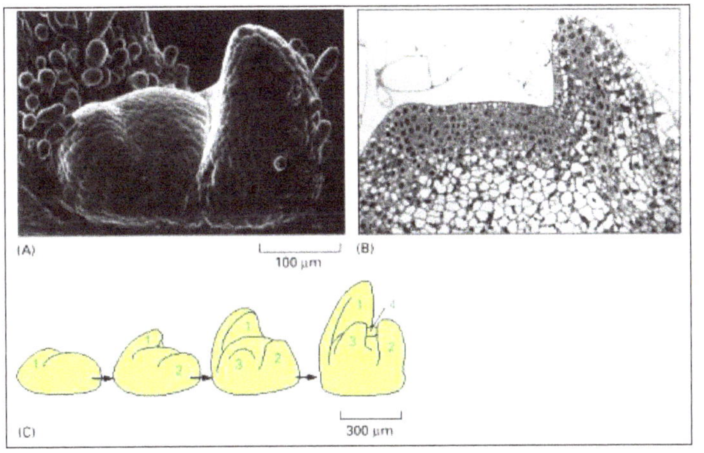

Shoot apex from a young tobacco plant

In figure, (A) A scanning electron micrograph shows the shoot apex with two sequentially emerging leaf primordia, seen here as lateral swellings on either side of the domed apical meristem. (B) A thin section of a similar apex shows that the youngest leaf primordium arises from a small group of cells (about 100) in the outer four or five layers of cells. (C) A very schematic drawing showing that the sequential appearance of leaf primordia takes place over a small distance and very early in shoot development. Growth of the apex will eventually form internodes that will separate the leaves in order along the stem.

Variations on this basic repetitive theme can give rise to more complex architectures, including structures such as tendrils, thorns, branches, and flowers. Thus, by switching on different sets of genes at the shoot apex, the plant can produce different types of primordia, in different spatial patterns.

Cell Signaling Maintains the Meristem

Central to all these phenomena is the question of how the apical meristem maintains itself. The meristem cells must continue to proliferate for weeks, years, or even centuries as a plant grows, replacing themselves while continually generating progeny cells that differentiate. Through all this, the size of the cluster of cells that constitute the meristem remains practically constant (about 100 cells in Arabidopsis, for example). New meristems may arise as the plant branches, but they too preserve the same size.

Genetic screens have identified genes required for meristem maintenance. For example, mutations that disrupt the WUSCHEL gene, which codes for a homeodomain protein, convert the apical meristem into non-meristematic tissue, so that the seedling fails to sprout. Conversely, mutations in the CLAVATA group of genes, coding for components of a cell-cell signaling pathway, make the meristem abnormally big. These genes are expressed in different layers of cells in the meristem region. The two most superficial cell layers, called the L1 and L2 layers, together with the uppermost part of the L3 layer, contain the cells of the meristem proper, capable of dividing indefinitely to give rise to future parts of the plant. The meristematic cells of the L1 and L2 layers express Clavata3, a small secreted signal protein. Just beneath, in the L3 layer, lies a cluster of cells expressing Clavata1 (the receptor for Clavata3). In the center of this Clavata1 patch are cells that express the Wuschel gene regulatory protein.

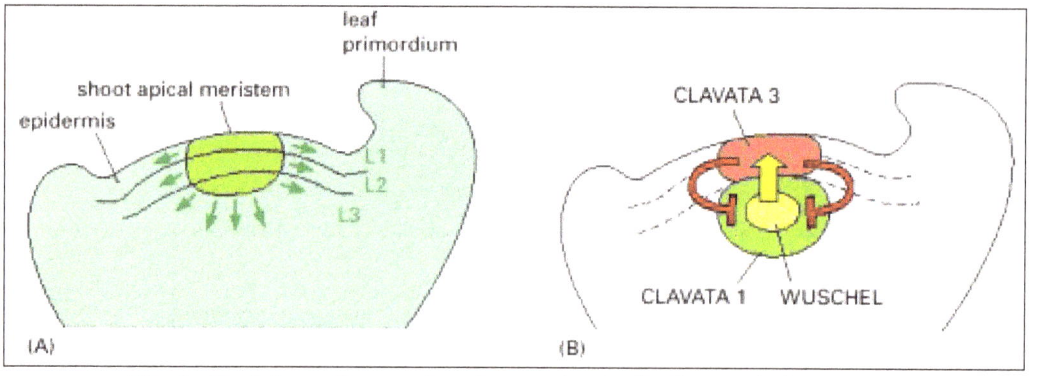

The feedback loops that are thought to maintain the shoot apical meristem

In figure, (A) The arrangement of cell layers constituting a shoot apical meristem. (B) The pattern of cell-cell communication that maintains the meristem. Artificial overexpression of Wuschel in the L3 region causes an increase in the number of cells in the L1 and L2 layers that behave as meristem cells and express Clavata3; artificial overexpression of Clavata3 in the L1 and L2 layers causes a reduction of Wuschel expression in the L3 region below and a decrease in the number of meristem cells. Clavata3 codes for a small signal protein, while Clavata1 codes for its receptor, a transmembrane protein kinase. Wuschel, which is expressed in the central part of the region that expresses the receptor Clavata1, codes for a gene regulatory protein of the homeodomain class. The size of the meristem is thought to be controlled by a self-regulating balance between a short-range stimulatory signal produced by cells expressing Wuschel (yellow arrow), and a longer-range inhibitory signal delivered by Clavata3 (red bars).

The pattern of cell divisions implies that the cells expressing Wuschel are not themselves part of the meristem proper; new Wuschel-expressing cells are apparently continually recruited from the meristematic part of the L3 population, just above the Wuschel domain. Nevertheless, the Wuschel-expressing cells are at the heart of the mechanism that maintains the meristem. A signal that they produce maintains meristematic behavior in the cells above, stimulates expression of the CLAVATA genes, and, presumably, causes new cells recruited into the Wuschel domain to switch on Wuschel. Negative feedback from the upper meristematic cells, delivered by the Clavata signaling pathway, acts back on the regions below to limit the size of the Wuschel domain, thereby preventing the meristem from becoming too big.

This account of the plant meristem is still uncertain in many details and other genes besides those we have mentioned are also involved. Nevertheless, mathematical modeling shows that systems of a similar sort, based on a feedback loop involving a short-range activating signal and a long-range inhibitory signal, can stably maintain a signaling center of a well-defined size even when there is continual proliferation and turnover of the cells that form that center. Analogous systems of signals are thought to operate in animal development to maintain localized signaling centers—such as the Organizer of the amphibian gastrula, or the zone of polarizing activity in a limb bud.

It is still not known how the Wuschel-expressing cells signal to their neighbors. One possibility is that the Wuschel protein itself diffuses directly from cell to cell through plasmodesmata—a signaling pathway peculiar to plants. Some other gene regulatory proteins have in fact been shown to travel in this way in meristems, spreading from cells that contain the corresponding mRNA into neighboring cells that do not.

- Regulatory Mutations can Transform Plant Topology by Altering Cell Behavior in the Meristem.

If a stem is to branch, new shoot apical meristems must be created, and this too depends on events in the neighborhood of the shoot apex. At each developing node, in the acute angle (the axil) between the leaf primordium and the stem, a bud is formed. This contains a nest of cells, derived from the apical meristem, that keep a meristematic character. They have the capacity to become the apical meristem of a new branch or the primordium of a structure such as a flower; but they also have the alternative option of remaining quiescent as axillary buds. The plant's pattern of branching is regulated through this choice of fate, and mutations that affect it can transform the structure of the plant. Maize provides a beautiful example.

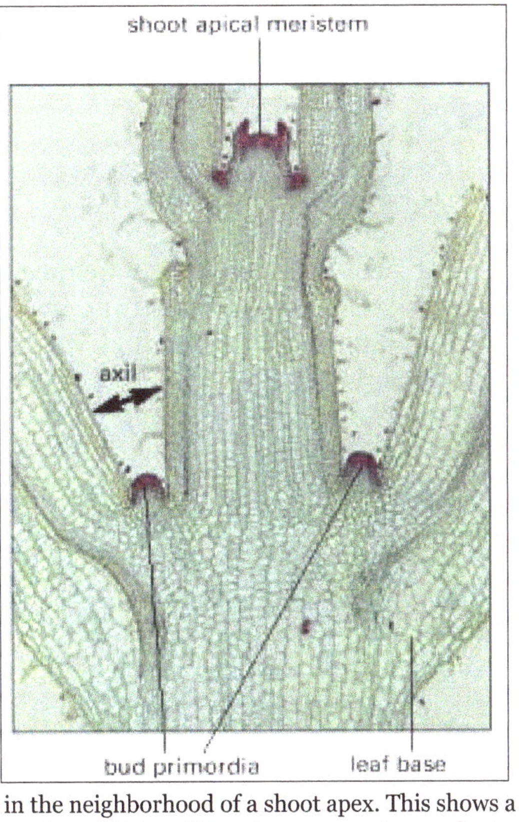

Axillary buds in the neighborhood of a shoot apex. This shows a longitudinal section of Coleus blumei, a common houseplant.

Maize represents one of mankind's most remarkable feats of genetic engineering. Native Americans created it by selective breeding, over a period of several centuries or perhaps millennia between 5,000 and 10,000 years ago. They started from a wild grass known as teosinte, with highly branched leafy stems and tiny ears bearing inedible hard kernels. Detailed genetic analysis has identified a handful of genetic loci—about five—as the sites of the mutations that account for most of the difference between this unpromising ancestor and modern corn. One of these loci, with a particularly dramatic effect, corresponds to a gene called teosinte branched-1 (tb1). In maize with loss-of-function mutations in tb1, the usual simple unbranched stem, with a few large leaves at intervals along it, is transformed into a dense, branching, leafy mass reminiscent of teosinte. The pattern of branching in the mutant implies that axillary buds, originating in normal positions, have escaped from an inhibition that prevents them, in normal maize, from growing into branches.

Transformation of plant architecture by mutation: a comparison of teosinte, normal maize, and tb1-defective maize.

In figure, (A) Photographs of three types of plants. (B) The architecture of teosinte, normal maize and the tb1-defective maize compared schematically. The tb1 gene product is needed for development of ears. It is absent in the tb1 mutant; it is present in both teosinte and normal maize, but these two plants differ because the gene is differently regulated.

In normal maize, the single stem is crowned with a tassel—a male flower—while a few of the axillary buds along the stem develop into female flowers and, upon fertilization, form the ears of corn that we eat. In the mutant maize with a defective tb1 gene, these fruitful axillary buds are transformed into branches bearing tassels. The wild teosinte plant is like the tb1-defective maize in its leafy, highly branched appearance, but unlike this mutant it makes ears on many of its side branches, as though tb1 were active. DNA analysis reveals the explanation. Both teosinte and normal maize possess a functional tb1 gene, with an almost identical coding sequence, but in maize the regulatory region has undergone a mutation that boosts the level of gene expression. Thus in normal maize the gene is expressed at a high level in every axillary bud, inhibiting branch formation, while in teosinte the expression in many axillary buds is low, so that branches are permitted to form.

This example shows how simple mutations, by switching the behavior of meri-stem cells, can transform plant structure—a principle of enormous importance in the breeding of plants for food. More generally, the case of tb1 illustrates how new body plans, whether of plant or animal, can evolve through changes in regulatory DNA without change in the characters of the proteins made.

Developmental Events in Separate Parts of the Plant

The fate of an axillary bud is dictated not only by its genes, but also by environmental conditions. Separate parts of a plant experience different environments and react to them individually by changes in their mode of development. The plant, however, must continue to function as a whole. This demands that developmental choices and events in one part of the plant affect developmental choices elsewhere. There must be long-range signals to bring about such coordination.

As gardeners know, for example, by pinching off the tip of a branch one can stimulate side growth: removal of the apical meristem relieves the quiescent axillary meristems of an inhibition and allows them to form new twigs. In this case the long-range signal from the apical meristem, or at least a key component has been identified. It is an auxin, a member of one of six known classes of plant growth regulators (sometimes called plant hormones), all of which have powerful influences on plant development. The five other known classes are the gibberellins, the cytokinins, abscisic acid, the gas ethylene, and the brassinosteroids. As shown in figure, all are small molecules that readily penetrate cell walls. They are all synthesized by most plant cells and can either act locally or be transported to influence target cells at a distance. Auxin, for example, is transported from cell to cell at a rate of about 1 cm per hour from the tip of a shoot toward its base. Each growth regulator has multiple effects, and these are modulated by the other growth regulators, as well as by environmental cues and nutritional status. Thus auxin alone can promote root formation, but in conjunction with gibberellin it can promote stem elongation, with cytokinin, auxin it can suppress lateral shoot outgrowth, and with ethylene it can stimulate lateral root growth.

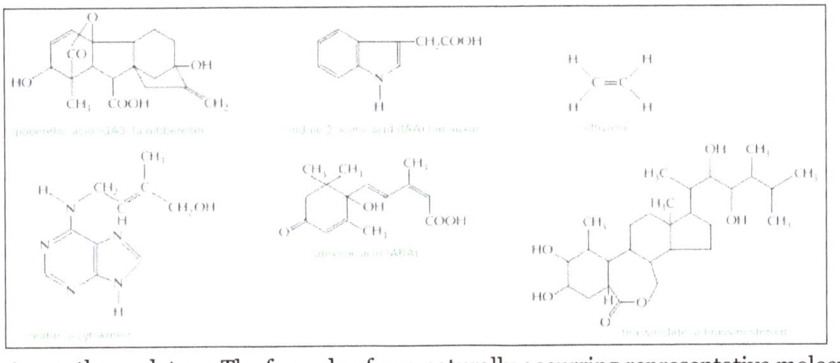

Plant growth regulators. The formula of one naturally occurring representative molecule from each of the six groups of plant growth regulatory molecules is shown.

Homeotic Selector Genes Specify the Parts of a Flower

Meristems face other developmental choices besides that between quiescence and growth, as we have already seen in our discussion of maize, and these also are frequently regulated by the environment. The most important is the decision to form a flower.

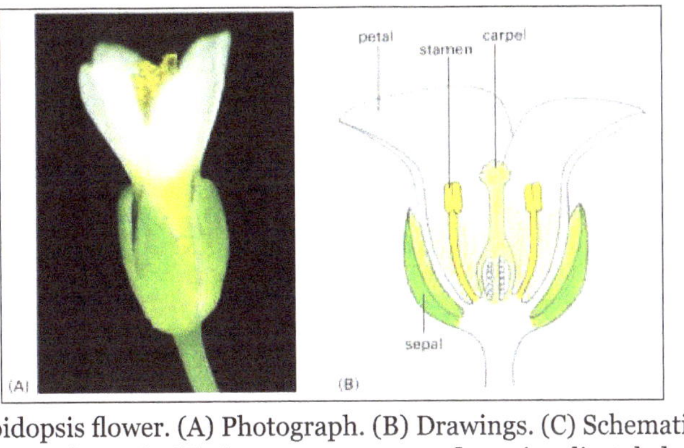

The structure of an Arabidopsis flower. (A) Photograph. (B) Drawings. (C) Schematic cross-sectional view. The basic plan, as shown in (C), is common to most flowering dicotyledonous plants.

The switch from meristematic growth to flower formation is typically triggered by light. By poorly understood mechanisms based on light absorption by phytochrome and cryptochrome proteins the plant can sense very precisely a change in day length. It responds by turning on expression of a set of floral meristem-identity genes in the apical meristem. By switching on these genes, the apical meristem abandons its chances of continuing vegetative growth and gambles its future on the production of gametes. Its cells embark on a strictly finite program of growth and differentiation: by a modification of the ordinary mechanisms for generating leaves, a series of whorls of specialized appendages are formed in a precise order—typically sepals first, then petals, then stamens carrying anthers containing pollen, and lastly carpels containing eggs. By the end of this process the meristem has disappeared, but among its progeny it has created germ cells.

The series of modified leaves forming a flower can be compared to the series of body segments forming a fly. In plants, as in flies, one can find homeotic mutations that convert one part of the pattern to the character of another. The mutant phenotypes can be grouped into at least four classes, in which different but overlapping sets of organs are altered. The first or 'A' class, exemplified by the apetala2 mutant of Arabidopsis, has its two outermost whorls transformed: the sepals are converted into carpels and the petals into stamens. The second or 'B' class, exemplified by apetala3, has its two middle whorls transformed: the petals are converted into sepals and the stamens into carpels. The third or 'C' class, exemplified by agamous, has its two innermost whorls transformed, with a more drastic consequence: the stamens are converted into petals, the carpels are missing, and in their place the central cells of the flower behave as a floral meristem, which begins the developmental performance all over again, generating another abnormal set of sepals and petals nested inside the first and, potentially, another nested inside that, and so on, indefinitely. A fourth class, the sepallata mutants, has its three inner whorls all transformed into sepals.

Arabidopsis flowers showing a selection of homeotic mutations.

In figure, (A) In apetala2, sepals are converted into carpels and petals into stamens; (B) In apetala3, petals are converted into sepals and stamens into carpels; (C) In agamous, stamens are converted into petals and carpels into floral meristem. (D) In a triple mutant where these three functions are defective, all the organs of the flower are converted into leaves.

These phenotypes identify four classes of homeotic selector genes, which, like the homeotic selector genes of Drosophila, all code for gene regulatory proteins. These are expressed in different domains and define the differences of cell state that give the different parts of a normal flower their different characters, as shown in figure The gene products collaborate to form protein complexes that drive expression of the appropriate downstream genes. In a triple mutant where the A, B and C genetic functions are all absent, one obtains in place of a flower an indefinite succession of tightly nested leaves. Conversely, in a transgenic plant where genes of the A, B and sepallata classes are all expressed together outside their normal domains, leaves are transformed into petals. Leaves therefore represent a "ground state" in which none of these homeotic selector genes are expressed, while the other types of organ result from expressing the genes in different combinations.

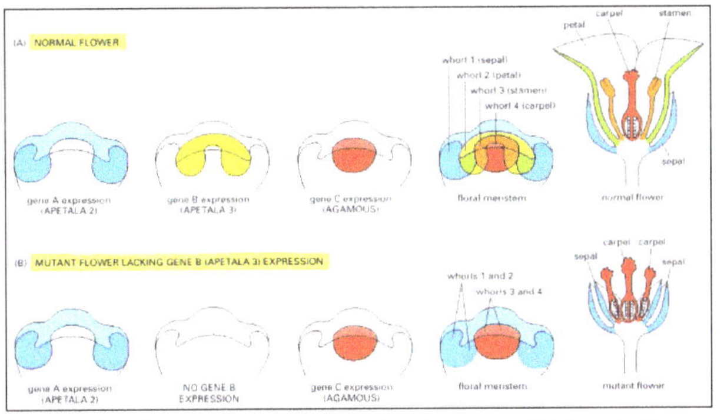

Homeotic selector gene expression in an Arabidopsis flower.

In figure, (A) Diagram of the normal expression patterns of the three genes whose mutant phenotypes are illustrated in Figure. All three genes code for gene regulatory proteins. The colored shading on the flower indicates which organ develops from each whorl of the meristem, but does not imply that the homeotic selector genes are still expressed at this stage. (B) The patterns in a mutant where the apetala3 gene is defective. Because the character of the organs in each whorl is defined by the set of homeotic selector genes that they express, the stamens and petals are converted into sepals and carpels. The consequence of a deficiency of a gene of class A, such as apetala2, is slightly more complex: the absence of this class A gene product allows the class C gene to be expressed in the outer two whorls as well as the inner two, causing these outer whorls to develop as carpels and stamens, respectively. Deficiency of a class C gene prevents the central region from undergoing terminal differentiation as a carpel and causes it instead to continue growth as a meristem, generating more and more sepals and petals.

Similar studies have been carried out in other plant species, and a similar set of phenotypes and genes have been identified: plants, no less than animals, have conserved their homeotic selector gene systems. Gene duplication has played a large part in the evolution of these genes: several of them, required in different organs of the flower, have clearly homologous sequences. These are not of the homeobox class but are members of another family of gene regulatory proteins (the so-called MADS family), also found in yeast and in vertebrates.

Clearly, plants and animals, have independently found very similar solutions to many of the fundamental problems of multicellular development.

Plants Growth and Development

Growth can be defined as an irreversible permanent increase in size of an organ or its parts or even of an individual cell. Generally, growth is accompanied by metabolic processes.

The root apical meristem and the shoot apical meristem provide the primary growth of the plants and also help to the elongation of the plants along their axis. The lateral meristems, vascular cambium and cork-cambium appear later m life in dicotyledonous plants and gymnosperms; these are the Meristems cause the increase in the girth of the organs in which they are active. This is a secondary growth of the plant as shown in figure.

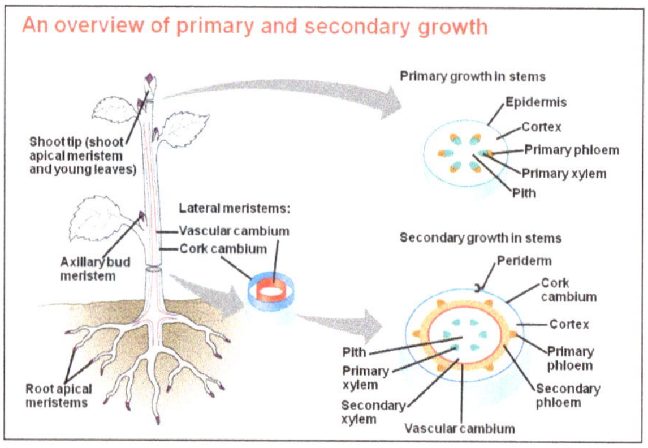

Diagrammatic representation of locations of root apical meristem, shoot aplical meristem and vascular cambium.

Locations of root apical meristem shoot apical meristem and vascular cambium. Arrows exhibit the direction of growth of cells and organ. The period of growth has three phases, meristematic, elongation and maturation as shown in figure above.

The constantly dividing cells, both at the root apex and the shoot apex, represent the meristematic phase of growth. This region cells are rich in protoplasm, possess large conspicuous nuclei. Their cell walls are primary in nature, thin and cellulosic with abundant plasmodesmatal connections. The cells being proximal to the meristematic zone are the phase of elongation.

The growth rate shows an increase which may be arithmetic or geometrical, during stages embryo development show geometric and arithmetic phases as shown in figure.

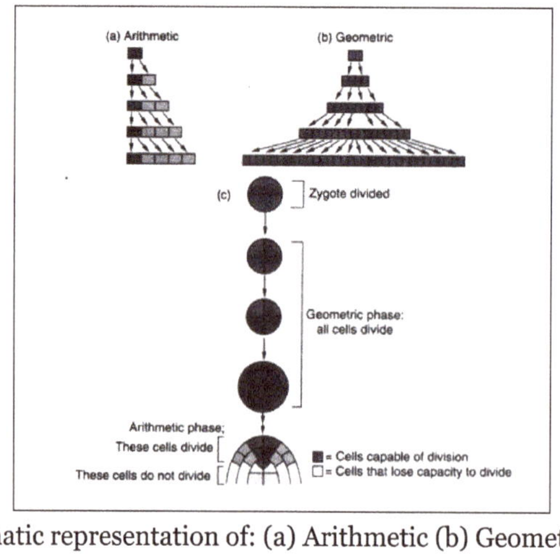

Diagrammatic representation of: (a) Arithmetic (b) Geometric growth

In figure shows the simplest expression of arithmetic growth is exemplified by a root elongating at a constant rate. If the length of the organ against time is plotted a linear curve is obtained. Initial by growth is slow and it increases rapidly thereafter- at an exponential rate. Here, both the progeny cells following mitotic cell division retain the ability to divide and continue to do so.

The growth slows down leading to a stationary phase due to limited nutrient supply. A typical sigmoid or S-curve affect plot the parameter of growth against time. This is shown in figure. A sigmoid curve is a characteristic of living organism growing in a natural environment.

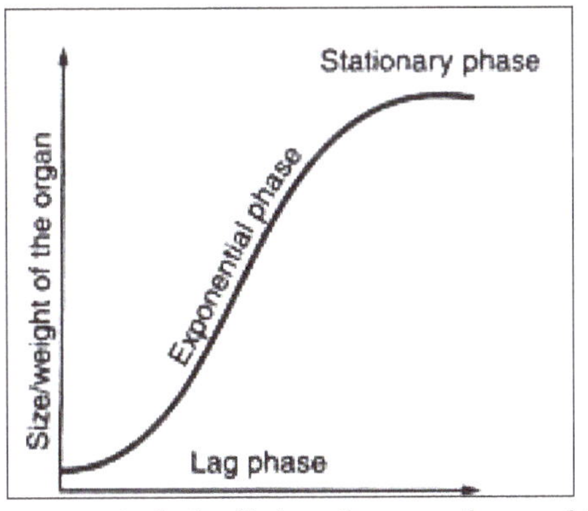

An idealised sigmoid growth curve typical of cells in culture, and many higher plants and plant organs

In figure two leaves, A and B, are drawn that are of different sizes but shows absolute increase in area in the given time to give leaves, A_1 and B_1. One of them shows much higher relative growth rate and it can be seen.

Water, oxygen and nutrients are very essential elements for growth. The plant cells, growth requires water. Turgidity of cells facilitates extension growth. Water also provides the medium for enzymatic activities needed for growth. Oxygen facilitates in releasing metabolic energy essential for growth activities. Nutrients are needed by plants for the synthesis of protoplasm.

The growth in plants is open. The final structure at maturity of a cell/tissue is also determined by the location of the cell within. The cells positioned away from root apical meristems differentiate as root-cap cells.

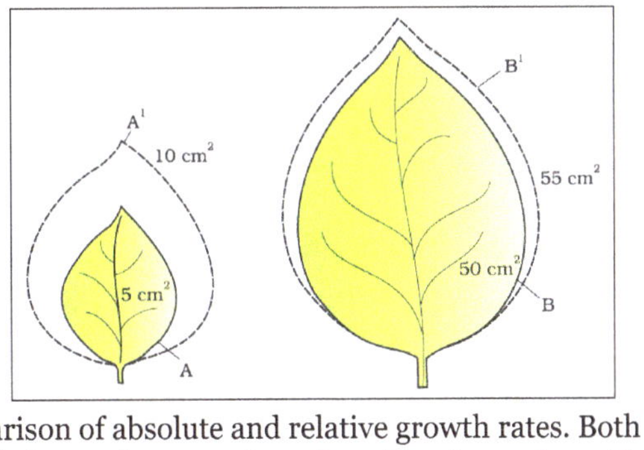

Diagrammatic comparison of absolute and relative growth rates. Both leaves A and B have increased their area by 5 cm2 in a given time to produce AI, B1 leaves.

Development

Development that includes all changes that an organism goes through during its life cycle from germination of the seed to senescence. Diagrammatic representation of the sequence of processes which constitute the development of a cell of a higher plant. It is also applicable to tissues/organs.

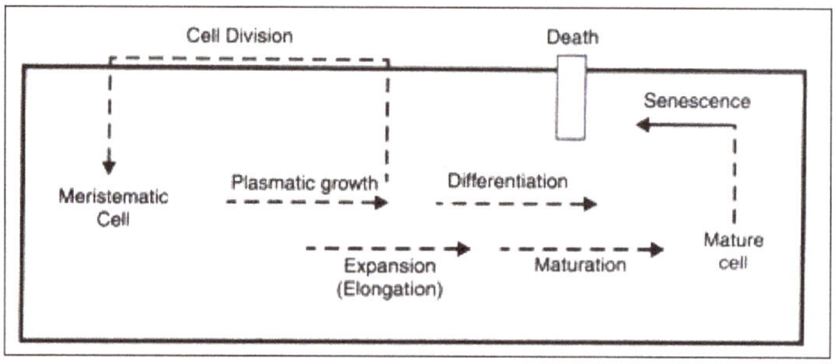

Sequence of the Developmental Process in a Plant Cell.

Plants follow different pathways in response to environment or phases of life to form different kinds of structures. In such plants, the leaves of the juvenile plant are different in shape from those in mature plants. Difference in shapes of leaves produced in air and those produced in water in buttercup also represent the heterophyllous development due to environment.

Growth, differentiation and development are very closely related events. Development is considered as the sum of growth and differentiation. Development in plants is under the control of intrinsic and extrinsic factors. The intense factors includes both intracellular or intercellular factors and the exoteric factors includes light, temperature, water, oxygen, nutrition, etc.

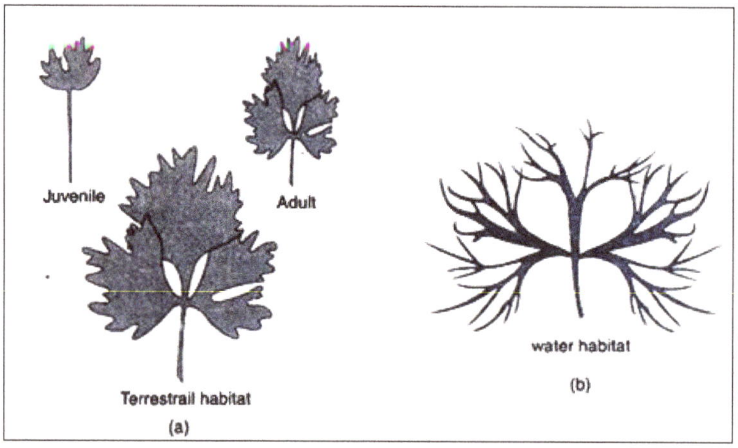

Heterophylly in (a) larkspur and (b) buttercup.

Plant Growth Regulators

The plant growth regulators (PGRs) are small, simple molecules of diverse chemical composition. The PGRs have two groups based on their functions. One group of PGRs is involved in growth promoting activities. The PGRs of the other group play an important role in plant responses to wounds and stresses of biotic and a biotic origin. The PGR abscise acid belongs to this group. The gaseous PGR, ethylene, could fit either of the groups.

Auxins as first isolated from human urine. The term 'auxin' is the indole-3- acetic acid (IAA), and to other natural and synthetic compounds having certain growth regulating properties. The growing apices of the stems and roots are produced by them. This is from where they migrate to the regions of their action. Axons like IAA and insole butyric acid (IBA) have been isolated from plants. NAA (naphthalene acetic acid) and 2, 4-D (2, 4-dichbrophenoxyacetic) are synthetic auxins.

The growing apical bud inhibits the growth of the lateral (axillary) buds in higher plants. Removal of shoot tips usually results in the growth of lateral buds as shown in figure. Tea plantations and hedge-making are the application.

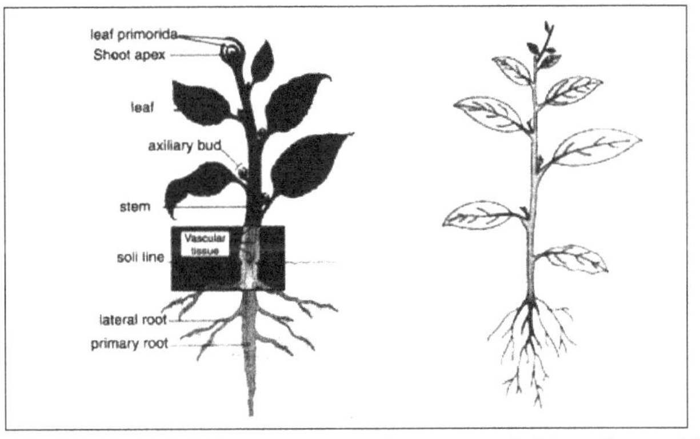

Apical dominance in plants: (a) A plant with apical bud intact and (b) A plant with apical bud removed

Spraying juvenile conifers with GAs hastens the maturity period, thus leading to early seed production. Gibberellins also promote bolting in beet, cabbages and many plants with rosette habit. Ethylene initiates flowering and for synchronizing fruit-set in pineapples, induces flowering in mango.

The most widely used compound as source of ethylene is ethephon. Ethephon in an aqueous solution is readily absorbed and transported within the plant and releases ethylene slowly. Ethephon hastens fruit ripening in tomatoes and apples and accelerates abscission in flowers and fruits.

Abscisic acid (ABA) has role in regulating abscission and dormancy. It also has other wide ranging effects on plant growth and development. It plays role of plant growth inhibitor and an inhibitor of plant metabolism. ABA inhibits seed germination. ABA helps seeds to withstand desiccation and other factors unfavorable for growth. ABA acts as an antagonist to GAs.

For any and every phase of growth, differentiation and development of plants, one or the other PGR has some role to play. Such roles could be complimentary or antagonistic. These could be individualistic or synergistic.

Similarly, there are a number of events in the life of a plant where more than one PGR interact to affect that event, e.g. for example, dormancy in seeds/buds, abscission, senescence, apical dominance, etc. They play an important role in plant growth and development.

Features of Plant Development

1. A single cell -the fertilized egg or zygote to multicellular.
2. Proliferation → embryo Organs and tissues begin to differentiate.
3. Formation of apical meristem early in embryonic development.
4. AM →primary body → secondary body (vascular and cork cambium).

Significance of this Mode of Development

Cell walls are rigid so cell replacement is impossible. Solution: continual additional new cells at the tips of roots and stems and lateral ones by secondary meristems.

Unique Features of Plant Development

1. Rigid cellulose walls -no cell and tissue migration. The three basic tissue systems (dermal, ground, and vascular), formed without gastrulation (animals) to establish this layered system of tissues. Plant cells divide and differentiate in place.
2. Germ cells are not set-aside early in development. This is also the case in several animal phyla, but it is always true for plants.
3. Plants have extended embryogenesis. Meristems persist long after embryogenesis.
4. Plasticity in development. One cell can regenerate whole plants. Compensates for the plant's lack of mobility.

Plant Life Cycle

1. Plant development alternates between haploid and diploid generations.
2. The embryo development is seen only in the diploid generation, but is itself produced by the fusion of gametes formed only by the haploid generation.
3. Therefore, the relationship between the two generations is important in the study of plant development.

Features of Plant Lifecycles

1. Exhibit alteration of generations-a diploid spore-producing generation (sporophyte) and a haploid gamete-producing generation (gametophyte).
2. Fertilization gives rise to a multicellular, diploid sporophyte that produces haploid spores via meiosis. Subsequent mitotic divisions within the gametophyte produce gametes.
3. The gametophytes and sporophytes of a species have distinct morphologies (they look different).
4. This above scheme seen in higher plants is a haplodiplontic life cycle ie Gametes are not the direct result of a meiotic division. Diploid sporophyte cells undergo meiosis to produce haploid spores. Each spore goes through mitotic divisions to yield a multicellular, haploid gametophyte.

Life Cycle of Flowering Plant

1. Develop a pollen tube that travels down the style to the ovary.
2. The generative nucleus undergoes mitotic division to give rise to two nuclei.

3. One sperm nucleus fuses with the egg cell to form a zygote and the other with the two polar nuclei to form an endosperm.
4. The zygote germinates to form the new sporophyte, which will eventually give rise to the gametophyte (pollen grains and embryo sac. The cycle repeats itself.

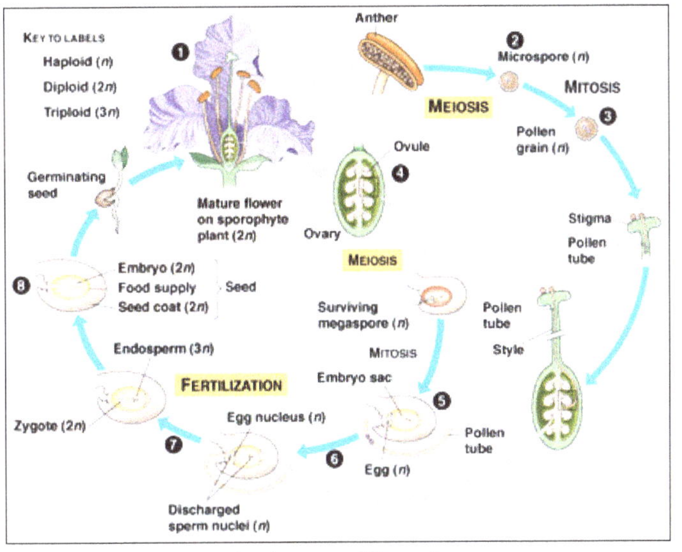

Angiosperm life cycle.

Gametogenesis -Creation of Gametes

General Male Development

Reproductive structure- stamen=microsporophyll has anthers containing microsporangia or lobes.

Anther structure

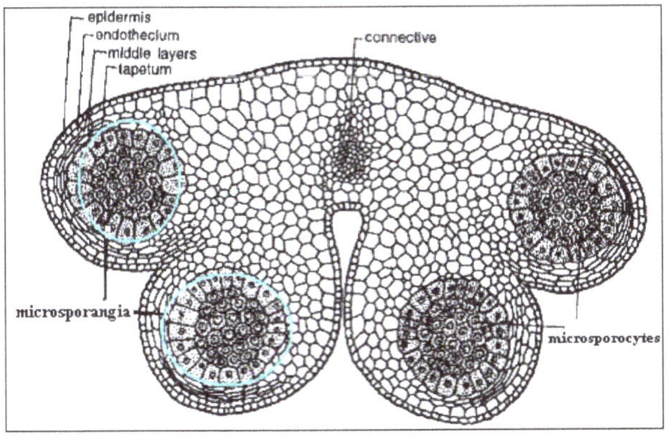

Structure of a young anther showing archesporium and sterile columella.

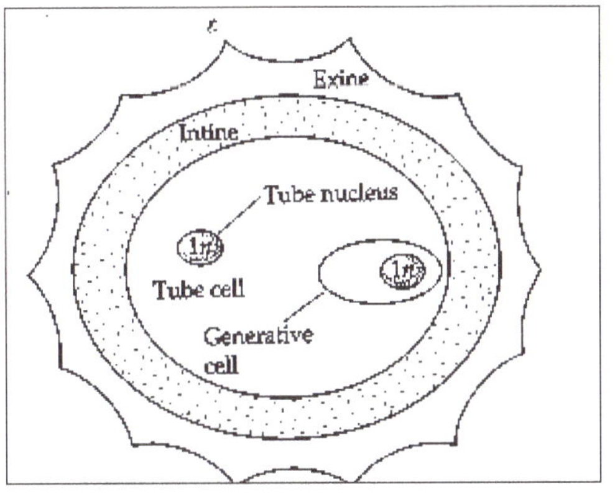

Pollen consists of a cell within a cell. The generative cell will undergo division to produce two sperm cells. One will fertilize the egg and the other will join with the polar nuclei, yielding the endosperm.

1. Anthers are typically tetrasporangiate ie 4-microsporangia. It has a column of sterile tissue called connective in the center of the two anther lobes. Each lobe has two microsporangia.

2. Each microsporangium is separated by a strip of sterile tissue. At maturity separation between the two sporangia in one lobe breaks down.

3. The anther later assumes a four-lobed shape. The hypodermal cells become prominent and constitute the archeosporium.

4. Parts of the anther:

- Anther wall- composed of:
    - Epidermis (function: protective, stimulates the tapetum),
    - Followed by the endothecium-2-3 layers (function; dehiscence of anthers at maturity) and
    - A single-layered tapetum- attains maximum development at tetrad sage of microspogenesis. Surrounds the sporogenous tissue.
- Sporogenous tissue- functions as microspore mother cells or microsporocytes or pollen mother cells (PMC).

5. Microgametogenesis is the process that is a microgametophyte developing in a pollen grain to the three-celled stage of its development. It begins with the formation of the microspores.

6. Each microsporocyte undergoes meiotic division to produce four haploid microspores (tetrad). They microspores separate soon after meiosis but some may remain tetrads and develop into compound pollen grains.

7. The microspore represents the beginning of mate gametophytic generation. Older microspores are referred to as pollen grains.

8. The microspores are eventually set free inside the anther locule by breakdown of callose wall that was produced by the tapetum.

9. The microspores synthesize individual walls. The callose plays a role in the laying down of the first pattern of exine.

10. Within the cytoplasm of each microspore, a microscopic, multicellular microgametophyte (male gametophyte) develops by mitosis. They become pollen once an outer exine is produced by the tapetum and the microspore) and the inner wall, the intine is produced by the microspore. Mature pollen consists of two cells within a cell.

11. The pollen tube is a unicellular outgrowth of the inner wall of the pollen grain and it is this tube, which transports the two sperm cells of the microgametophyte to the egg cell of the megagametophyte in the ovule.

12. The tube cell contains a generative cell within it.

- The generative cell divides to produce two sperm nuclei.
- The tube cell nucleus guides pollen germination and the growth of the pollen tube after the pollen lands on the stigma of the female parent.

- One of the sperm will fuse with the egg cell to produce the next sporophyte generation.
- The second sperm will fuse with the two polar nuclei to form the endosperm that provides nourishment for the embryo.

NB. Pollen is the immature male gametophyte while the germinated pollen grain, with its pollen tube and two sperm nuclei represents the mature male gametophyte.

General Female Development- Ovule

Ovule is composed of the:

1. Megasporangium or embryo sac enclosed by its protective coat the integuments. It is attached to the placenta on the inner wall of the ovary by a stalk called funiculus.
2. Below the integument is the Endothelium. Function-nutritive- has large concentrations of proteins, RNA, carbohydrates and enzymes.
3. Within the nucellus is the embryo sac or female gametophyte that houses the megasporocyte cell or megaspore mother cell.

Megasporogenesis

1. The megasporocyte cell undergoes meiotic division to produce four megaspores. The megaspores are retained in the ovule and never released from the megasporangium.
2. Generally, three of the four megaspores degenerate.
3. Within the cytoplasm of the surviving functional megaspore (endosporic development), a microscopic, multicellular megagametophyte (female gametophyte) develops after 3 rounds of mitosis divisions. It is an 8-celled structure.
    - One cell of this gametophyte phase will function as an egg. It is found at the micropylar end together with 2 synergid cells. Function: provide nutrition for the embryo.
    - At its chalazal end are 3 antipodal cells. Function: nutrition, storage and secretion of growth substances.
    - The egg cell, synergids and antipodal cell are uninucleate and haploid.
    - In the middle is the central cell or primary endosperm cell with 2 nuclei (polar nuclei). Function contains sufficient food reserves available during fertilization and endosperm development.
4. Concurrent with these events, a protective envelope of tissue, the integument, develops around the megasporangium of the ovule.
5. A narrow canal in this envelope, called the micropyle, provides an open pathway to the internal megasporangium and megagametophyte. It is the integument that develops into the protective seed coat of the mature seed.

The female gametophyte (embryo sac) is housed in the ovule.

Double Fertilization

1. After pollination the pollen tube grows through the stigma and style toward the ovules in the ovary.
2. The germ cell in the pollen grain divides and releases two sperm cells, which move down the pollen tube.
3. Once the tip of the tube reaches the micropyle end of the embryo sac, the tube grows through into the embryo sac through one of the synergids, which flank the egg.
4. One sperm cell fuses with the egg, producing the zygote that will later develops into the next-generation sporophyte.
5. The second sperm cell fuses with the two polar bodies located in the center of the sac, producing the nutritive triploid endosperm tissue that will provide nutrition for the embryo's growth and development.
6. The embryo sac degenerates after fertilization.

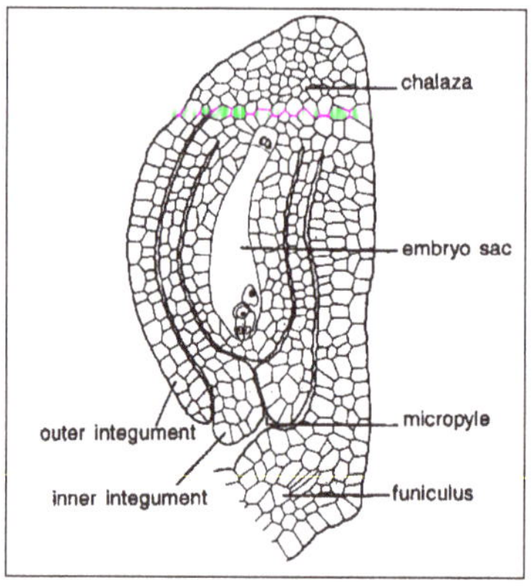

Longitudinal section of an angiosperm ovule

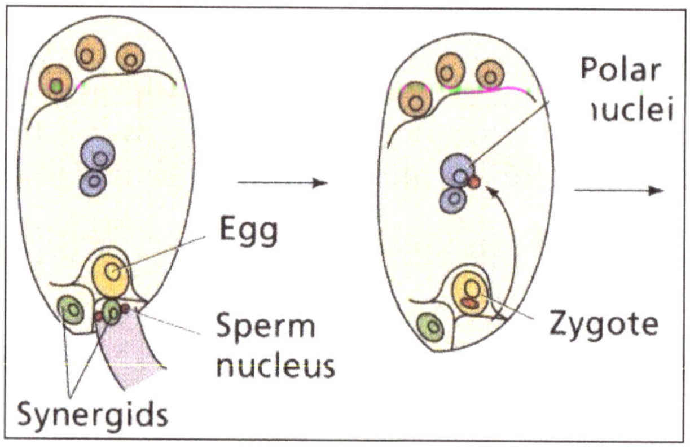

Showing the events of "double fertilization" of the egg and polar nuclei by the two sperm cells.

Stages of Plant Growth

First Stage – Seed Germination

The plant's life cycle starts with germination of the seed. All you need is water, oxygen, and the right temperature for the seed to germinate. How about light? Well, most seeds are not affected by light or darkness, but some seeds will not germinate until they see sufficient light.

As the right temperature is reached, and water is provided, the seed will germinate. When a dry seed uptakes water, we call it imbibition. Imbibition means to drink; seeds imbibe water.

This process is so similar to a dry sponge exposed to water. As the sponge absorbs water, it grows in size. The color of the sponge becomes more vivant and becomes heavier. Seeds do the same.

Right Temperature for Germination

Most common annual vegetables have optimal germination temperatures between 75-89 °F (24-32 °C). Some species like radishes or spinach can germinate at significantly lower temperatures, as low as 40 °F (4 °C).

Swelling to grow

As the seed drinks water, it expands while enzymes and food supplies in the seed become hydrated. The seed is now activated and starts to go to work. The increased metabolic activity provides energy for the growth process. In addition, the water causes turgor pressure to increase in the cells.

Turgor pressure is the outward pressure that occurs in a plant cell when the seed's cytoplasm and vacuoles fill up with water, and the cell membrane pushes against the cell walls. These guys swell up like water balloons in cells of the seed.

Time to make Roots

Roots start to emerge from the seed. This activity is the first indication that the seed is viable and will grow. Eventually, the plant shoot, top of the plant, will sprout out and emerge from the seed.

Attached to the shoot are tiny leaves. Once these leaves begin to manufacture food, the plant will grow and enter the seedling stage.

Plant will now develop its first set of true leaves and roots.

The root growth that takes place at this time is crucial to how fast the plant will grow. Providing the ideal environment for the roots will ensure that your crop will have a chance to flourish.

Darkness

If germination occurs in darkness, root growth slows after the shoot emerges and shoot elongation accelerates. This behavior occurs because the seed thinks it's still covered with soil and needs to emerge from soil to see the light where it can obtain energy from the sunlight by photosynthesis. Once it is in the light, the plant undergoes dramatic changes like turning green and producing leaves. This transformation with the light is called photo-morphogenesis.

Second Stage – Vegetative and Growth

Now that the root system can support growth, the vegetative stage begins. Plant nutrient requirements call for a large amount of nitrogen for the production of chlorophyll. Growth during this stage is primarily dedicated to stem, branch, and leaf areas.

The most substantial growth over the life cycle of the plant occurs in the vegetative stage and will continue unless hindered by a change in environment or lack of water and nutrients.

Third Stage – Reproductive, Flowering and Fruit

The final stage of the plant growth is the reproductive stage. Since the objective is now to reproduce, the plant's energies are directed to the production of flowers, fruits, and seeds.

Nitrogen is no longer the main nutrient the plant needs. We now need to boost up other nutrients like phosphorus and potassium (P-K). This is all due to a slowing down of plant growth.

Bloom

This is when a change in nutrients is made, from vegetative formula to a flowering, or 'bloom' formula. This is why many aeroponics solution nutrients come as a two or three-part system.

In some plants, the reproductive stage is triggered by the change in the length of the daylight, this process is called photoperiodism. The length of the day is what governs plants grown outdoors to flower or produce fruit.

When growing indoors we must provide the right photoperiod for the crop to develop properly. Changing the length of artificial light can trick the plant to deliver flowers or fruits early. Commercial growers use this technique to provide flowers to markets out of season. As a reward, these growers demand a premium from vendors and customers alike.

Pollination – Spreading the Love

Plants are pollinated by various means like wind, water or bees and other insects. Pollination occurs when the pollen from one plant is transferred to another plant. Reproduction in flowering plants begins with pollination, the transfer of pollen from one plant to the stigma of another flower or to the stigma of another flower of the same plant.

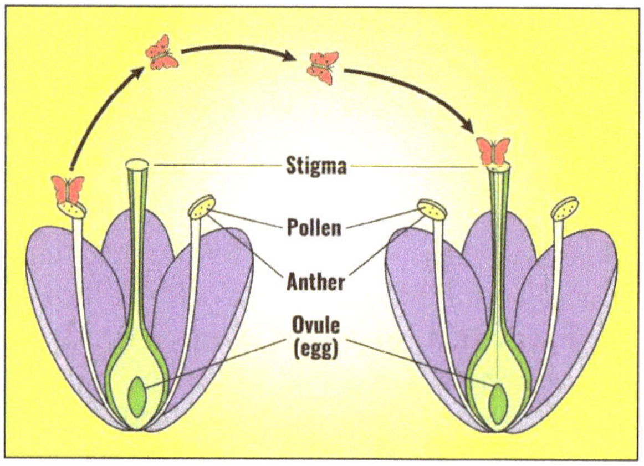

Pollination is required for the fertilization of the plant. Pollen contains the male gametes, or sperm, of the plant. The male gametes are transferred through this pollination process to the stigma of the plant. The stigma contains the female gamete (egg cell) where fertilization can occur with the male gamete.

When pollination occurs between plants of the same species it is called self-pollination. Pollination between plants of different species it is called cross-pollination.

Fertilization

Once the pollination is completed, a pollen tube grows into the ovary of the plant. The male gamete cells move along the pollen tube toward the ovary. Here the male gamete cell fertilizes the female gamete cell. This results in the development of a zygote. Zygotes develop into embryos and become the beginning basis of a new plant.

Now comes the seeds. A second male gamete cell fuses with polar cell bodies found in the center of the embryo sac, and produce the endosperm tissue material that provides food and energy for the development of the embryo. This endosperm is the seed or fruit produced by the plant. The fertilized ovule develops into a seed.

Bees

Now for growing indoors away from natural insects, you must play 'bee' by pollinating the flowers on mature plants manually. A delicate touch with a soft painter's brush to each flower is all is needed to pollinate the plant to produce fruit.

For larger crops a plant 'shaker' is used to vibrate the pollen loose to cross pollinate the plants. However, a good breeze from a high velocity fan is usually sufficient to pollinate indoor plants in a confined area. This be the method to use for our aeroponics system. We want to automate as much as possible for repeatable and transferrable success.

Plant Embryogenesis

Plant embryogenesis is a process that occurs after the fertilization of an ovule to produce a fully developed plant embryo. This is a pertinent stage in the plant life cycle that is followed by dormancy and germination. The zygote produced after fertilization, must undergo various cellular divisions and differentiations to become a mature embryo. An end stage embryo has five major components including the shoot apical meristem, hypocotyl, root meristem, root cap, and cotyledons. Unlike animal embryogenesis, plant embryogenesis results in an immature form of the plant, lacking most structures like leaves, stems, and reproductive structures.

Morphogenic Events

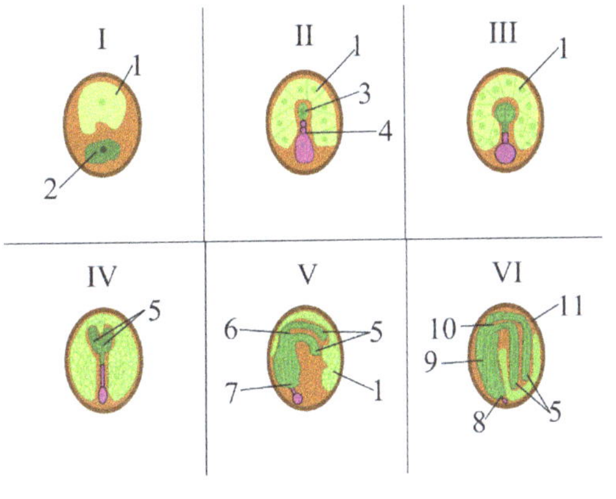

Embryogenesis occurs naturally as a result of single, or double fertilzation, of the ovule, giving rise to two distinct structures: the plant embryo and the endosperm which go on to develop into a seed. The zygote goes through various cellular differentiations and divisions in order to produce a mature embryo. These morphogenic events form the basic cellular pattern for the development of the shoot-root body and the primary tissue layers; it also programs the regions of meristematic tissue formation. The following morphogenic events are only particular to eudicots, and not monocots.

Table: Six moments in embryogenesis.

Two cell stage	endosperm
Eight cell stage	single celled zygote
Globular stage	embryo
Heart stage	suspensor
Torpedo stage	cotyledons
Maturation	shoot apical meristem (SAM) root apical meristem (RAM)

Two Cell Stage

The zygote and endosperm are present within the ovule. Then the zygote undergoes an asymmetric transverse cell division that gives rise to two cells - a small apical cell resting above a large basal cell. These two cells are very different, and give rise to different structures, establishing polarity in the embryo.

- Apical cell: The small apical cell is on the top and contains most of the cytoplasm, the aqueous substance found within cells, from the original zygote. It gives rise to the hypocotyl, shoot apical meristem, and cotyledons.
- Basal cell: The large basal cell is on the bottom and consists of a large vacuole and gives rise to the hypophysis and the suspensor.

Eight Cell Stage

After two rounds of longitudinal division, and one round of transverse division, an eight-celled embryo is the result. The embryo looks like during the eight cell stage. According to Laux et al., there are four distinct domains during the eight cell stage. The first two domains contribute to the embryo proper. The *apical embryo domain*, gives rise to the shoot apical meristem and cotyledons. The second domain, the *central embryo domain*, gives rise to the hypocotyl, root apical meristem, and parts of the cotyledons. The third domain, the *basal embryo domain,* contains the hypophysis. The hypophysis will later give rise to the radicle and the root cap. The last domain, the *suspensor*, is the region at the very bottom, which connects the embryo to the endosperm for nutritional purposes.

Sixteen Cell Stage

Additional cell divisions occur, which leads to the sixteen cell stage. The four domains are still present, but they are more defined with the presence of more cells. The important aspect of this stage is the introduction of the protoderm, which is meristematic tissue that will give rise to the epidermis. The protoderm is the outermost layer of cells in the embryo proper.

Globular Stage

The name of this stage is indicative of the embryo's appearance at this point in embryogenesis; it is spherical or globular. Stage III, in the photograph above, depicts what the embryo looks like during the globular stage. 1 is indicating the location of the endosperm. The important component of the globular phase is the introduction of the rest of the primary meristematic tissue. The protoderm was already introduced during the sixteen cell stage. According to Evert and Eichhorn, the ground meristem and procambium are initiated during the globular stage. The ground meristem will go on to form the ground tissue, which includes the pith and cortex. The procambium will eventually form the vascular tissue, which includes the xylem and phloem.

Heart Stage

According to Evert and Eichhorn, the heart stage is a transition period where the cotyledons finally start to form and elongate. It is given this name in eudicots because most plants from this group have two cotyledons, giving the embryo a heart shaped appearance. The shoot apical meristem is between the cotyledons.

Torpedo Stage

This stage is defined by the continued growth of the cotyledons and axis elongation. In addition, programmed cell death must occur during this stage. This is carried out throughout the entire growth process, like any other development. However, in the torpedo stage of development, parts of the suspensor complex must be terminated. The suspensor complex is shortened because at this point in development most of the nutrition from the endosperm has been utilized, and there must be space for the mature embryo. After the suspensor complex is gone, the embryo is fully developed. Stage V, indicates what the embryo looks like at this point in development.

Maturation

The second phase, or postembryonic development, involves the maturation of cells, which involves cell growth and the storage of macromolecules (such as oils, starches and proteins) required as a 'food and energy supply' during germination and seedling growth.

Dormancy

The end of embryogenesis is defined by an arrested development phase, or stop in growth. This phase usually coincides with a necessary component of growth called dormancy. Dormancy is a period in which a seed cannot germinate, even under optimal environmental conditions, until a specific requirement is met. Breaking dormancy, or finding the specific requirement of the seed, can be rather difficult. For example, a seed coat can be extremely thick. According to Evert and Eichhorn, very thick seed coats must undergo a process called scarification, in order to deteriorate the coating. In other cases, seeds must experience stratification. This process exposes the seed to certain environmental conditions, like cold or smoke, to break dormancy and initiate germination.

Role of Auxin

Auxin is a hormone related to the elongation and regulation of plants. It also plays an important role in the establishment polarity with the plant embryo. Research, conducted by Cooke, Racusen, and Cohen, has shown that the hypocotyl from both gymnosperms and angiosperms show auxin transport to the root end of the embryo They hypothesized that the embryonic pattern is regulated by the auxin transport mechanism, and the polar positioning of cells within the ovule. The importance of auxin was shown, in their research, when carrot embryos, at different stages, were subjected to auxin transport inhibitors. The inhibitors that these carrots were subjected to made them unable to progress to later stages of embryogenesis. During the globular stage of embryogenesis, the embryos continued spherical expansion. In addition, oblong embryos continued axial growth, without the introduction of cotyledons. During the heart embryo stage of development there were additional growth axes on hypocotyls. Further auxin transport inhibition research, conducted on *Brassica juncea*, show that after germination, the cotyledons were fused and not two separate structures.

Alternative Forms of Embryogenesis

Somatic Embryogenesis

Somatic embryos are formed from plant cells that are not normally involved in the development of embryos, i.e. ordinary plant tissue. No endosperm or seed coat is formed around a somatic embryo. Applications of this process include: clonal propagation of genetically uniform plant material; elimination of viruses; provision of source tissue for genetic transformation; generation of whole plants from single cells called protoplasts; development of synthetic seed technology. Cells derived from competent source tissue are cultured to form an undifferentiated mass of cells called a callus. Plant growth regulators in the tissue culture medium can be manipulated to induce callus formation and subsequently changed to induce embryos to form the callus. The ratio of different plant growth regulators required inducing callus or embryo formation varies with the type of plant. Asymmetrical cell division also seems to be important in the development of somatic embryos, and while failure to form the suspensor cell is lethal to zygotic embryos, it is not lethal for somatic embryos.

Androgenesis

The process of androgenesis allows a mature plant embryo to form from a reduced, or immature, pollen grain. Androgenesis usually occurs under stressful conditions. Embryos that result from this mechanism can germinate into fully functional plants. As mentioned, the embryo results from a single pollen grain. Pollen grains consists of three cells - one vegetative cell containg two generative cells. According to Maraschin et al., androgenesis must be triggered during the asymmetric division of microspores. However, once the vegetative cell starts to make starch and proteins, androgenesis can no longer occur. Maraschin et al., indicates that this mode of embryogenesis consists of three phases. The first phase is the *acquisition of embryonic potential*, which is the repression of gametophyte formation, so that the differentiation of cells can occur. Then during the *initiation of cell divisions*, multicellular structures begin to form, which are contained by the exine wall. The last step of androgenesis is *pattern formation*, where the embryo-like structures are released out of the exile wall, in order for pattern formation to continue.

After these three phases occur, the rest of the process falls in line with the standard embryogenesis events.

Plant Growth and Buds

Embryonic tissue is made up of actively growing cells and the term is normally used to describe the early formation of tissue in the first stages of growth. It can refer to different stages of the sporophyte and gametophyte plant; including the growth of embryos in seedlings, and to meristematic tissues, which are in a persistently embryonic state, to the growth of new buds on stems.

In both gymnosperms and angiosperms, the young plant contained in the seed, begins as a developing egg-cell formed after fertilization (sometimes without fertilization in a process called apomixis) and becomes a plant embryo. This embryonic condition also occurs in the buds that form on stems. The buds have tissue that has differentiated but not grown into complete structures. They can be in a resting state, lying dormant over winter or when conditions are dry, and then commence growth when conditions become suitable. Before they start growing into stem, leaves, or flowers, the buds are said to be in an embryonic state.

Sporophyte

Sporophyte means "spore-bearing" plant, is the diploid multicellular phase of an organism that displays alternation of generations. The sporophyte phase develops from the fertilized egg, or zygote, by simple cell division and subsequent differentiation. Sporophytes occur in a few algae and aquatic fungi, and are universal in true plants. They differ greatly in size and level of complexity. In bryophytes the sporophyte is short-lived and permanently attached to the female parent, upon which it is nutritionally dependent. In all other plants, the sporophyte becomes independent of the female parent soon after embryological development is completed and remains as the dominant, photosynthetic stage of the plant. In the simplest case, a sporophyte can consist of only a capsule, or sporangium, as in the liverwort Riccia, but usually it will also possess one or more vegetative organs. For example, in mosses the sporophyte consists of a foot; a green, stemlike seta; and a single complex sporangium, while in pines it is a highly branched tree with roots, stems, leaves and thousands of sporangia. Within the sporangia, which are the reproductive organs of the sporophyte, haploid spores are produced by meiosis. Germination of these spores marks the beginning of the haploid, sexual phase of the life cycle.

Cleistocarpous sporophyte of the moss Physcomitrella patens.

Sporophyte, in certain plants and algae, nonsexual phase (or an individual representing the phase) in the alternation of generations—a phenomenon in which two distinct phases (a haploid and a diploid phase) occur in the life history of a plant, each phase producing the other. (The alternate, sexual phase is the gametophyte). In the sporophyte phase, a diploid plant body grows and eventually produces spores through meiosis. These spores divide mitotically to produce haploid, gamete-producing bodies called gametophytes. The union of two gametes during fertilization produces a diploid zygote, which divides mitotically to form a new sporophyte.

The character and relative extent of the two phases vary greatly among different groups of plants and algae. During the course of evolution, the gametophyte stage is progressively reduced; thus, in the higher (i.e., vascular) plants, the sporophyte is the dominant phase in the life cycle, whereas in the more primitive nonvascular plants the gametophyte remains dominant.

All land plants, and some algae, have life cycles in which a haploid gametophyte generation alternates with a diploid sporophyte, the generation of a plant or alga that has a double set of chromosomes. A multicellular sporophyte generation or phase is present in the life cycle of all land plants and in some green algae. For common flowering plants (Angiosperms), the sporophyte generation comprises almost their whole life cycle (i.e. whole green plant, roots etc), except phases of small reproductive structures (pollen and ovule).

Young sporophytes of the common moss Tortula muralis. In mosses, the gametophyte is the dominant generation, while the sporophytes consist of sporangium-bearing stalks growing from the tips of the gametophytes.

The sporophyte produces spores (hence the name), by meiosis. These meiospores develop into a gametophyte. Both the spores and the resulting gametophyte are haploid, meaning they only have one set of homologous chromosomes. The mature gametophyte produces male or female gametes (or both) by mitosis. The fusion of male and female gametes produces a diploid zygote which develops into a new sporophyte. This cycle is known as alternation of generations or alternation of phases.

In the normal course of events, the zygote and sporophyte will have a full double set of chromosomes again. An exception is when a diploid and haploid gamete fuse, resulting in a triploid sporophyte, which will usually be sterile, as dividing three sets of chromosomes into two halves causes complications.

In flowering plants, the sporophyte comprises the whole multicellular body except the pollen and embryo sac.

Bryophytes (mosses, liverworts and hornworts) have a dominant gametophyte stage on which the adult sporophyte is dependent on the gametophyte for nutrition. The embryo of the sporophyte develops from the zygote within the female sex organ or archegonium, and in its early development is therefore nurtured by the gametophyte. Because this embryo-nurturing feature of the life cycle is common to all land plants they are known collectively as the Embryophytes. Most algae have dominant gametophyte generations, but in some species the gametophytes and sporophytes are morphologically similar (isomorphic). An independent sporophyte is the dominant form in all clubmosses, horsetails, ferns, gymnosperms, and angiosperms (flowering plants) that have survived to the present day.

Early land plants had sporophytes that produced identical spores (isosporous or homosporous) but the ancestors of the gymnosperms evolved complex heterosporous life cycles in which the spores producing male and female gametophytes were of different sizes, the female megaspores tending to be larger, and fewer in number, than the male microspores.

During the Devonian period several plant groups independently evolved heterospory and subsequently the habit of endospory, in which single megaspores were retained within the sporangia of the parent sporophyte, instead of being freely liberated into the environment as in ancestral exosporous plants. These endosporic megaspores contained within them a miniature multicellular female gametophyte complete with female sex organs or archegonia containing oocytes which were fertilised by free-swimming sperm produced by windborne miniatuarised male gametophytes in the form of pre-pollen. The resulting zygote developed into the next sporophyte generation while still retained within the pre-ovule, the single large female meiospore or megaspore contained in the modified sporangium or nucellus of the parent sporophyte. The evolution of heterospory and endospory were among the earliest steps in the evolution of seeds of the kind produced by gymnosperms and angiosperms today.

Root System

A root is the part of the plant responsible for anchoring it down to the ground and absorbing essential nutrients, minerals and water from the soil. It also used to store food. But not all plants have their roots underground, for instance, plants such as ivy have roots that are present entirely off the ground.

Types of Roots

All roots have similar functions, however, their structure varies. Hence, based on these criteria, roots are classified into two types:

- Tap-root system: Taproots have a main central root upon which, small, lateral roots called root hairs are attached. Mustard, Carrot, Beetroot, Parsley, China rose, etc. are some examples of Taproots system.
- Fibrous root system: Fibrous roots, on the other hand, are bushy roots in which thin, moderately branching roots grow from the stem. Rice, Wheat, Maize, Marigold, Banana, etc.are some examples of Fibrous system.

Functions of the Root

Roots perform various functions that are necessary for the survival of the plants. They are an Integral or Integrated system that helps the plant in:

- Anchoring: Roots are the reason plants remain attached to the ground. They support the plant body, ensuring that it stands erect.
- Absorption: Primary function of roots is to absorb water and dissolved minerals from the soil. This is crucial as it helps in the process of photosynthesis.
- Storage: Plants prepares food and store in the form of starch in the leaves, shoots and roots. Prominent examples include carrots, radish, beetroot, etc.
- Reproduction: Even though roots are not the reproductive part of plants, they are vegetative parts. In some plants, roots are a means of reproduction. For instance, new plants arise from creeping horizontal stems called runners (stolons) in jasmine, grass, etc. This type of reproduction is called vegetative propagation.
- Ecological Function: They check soil erosion, provide sustenance and also habitat to various organisms.

Shoot System

The plant shoot system is a complex network of a number of different parts all working to keep the plant healthy and growing. It is fascinating how much is involved in what we usually look at as a simple living organism. We will investigate the shoot system here.

The Structure of the Shoot System

A plant has many complicated and complex systems that keep it living and growing, including the shoot system. When referring to the shoot system in a plant, we generally refer to the leaves, buds, flowering stems and flowering buds, as well as the main stem itself. The word 'shoot' generally is used when talking about the main stem.

As we move from the ground surface to the terminal bud (end of the undeveloped shoot) we will encounter nodes and internodes. Nodes are the points where leaves are attached, and internodes are the places on the stem between the nodes. In the crux created by the node and stem, there are axillary buds that lay dormant but have the potential to grow a vegetative branch.

These axillary buds lay dormant because of apical dominance. This is a phenomena where the plant concentrates most of its resources at the terminal bud. The terminal buds grows at the apex, or tip of the plant. This is responsible for making the plant grow taller and bigger so it makes sense that the plant would want to concentrate resources here. If the terminal bud gets damaged, the axillary buds will 'wake up' and begin to grow, saving the plant.

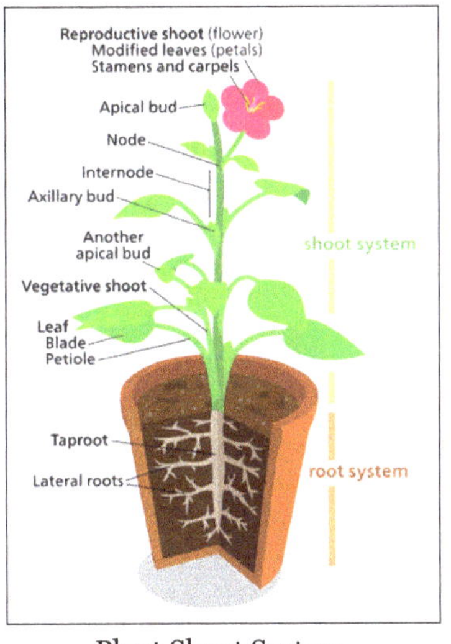

Plant Shoot System.

If the axillary buds begin to grow, they can create their own terminal buds, leaves, etc. They, in a sense, create a secondary plant themselves. If you own any houseplants, you might understand the idea of pruning them to create 'bushy' plants. Well, this is because you are removing the terminal bud that stimulates the axillary buds to grow, creating that thicker looking plant.

As is commonly known, the leaves of the plant act as the photosynthetic factory for it, producing sugars and other compounds for the plant to survive on. There are many different types of leaves, but most will include a blade, stalk, veins and the petiole. The stalk is the main stem of the leaf, the blade is the actual green leaf portion, and the petiole is where it attaches to the node.

The shoot system or the ascending axis developed out of the plumule while below, covered by the soil, and is an equally important portion— the root system.

The shoot is the emblem of greenness because of the green leaves while the root, growing in darkness, is never green.

The skeleton of the shoot is formed by the stem. The main stem is developed as a direct prolongation of the tigellum of the embryo and grows vertically up while its branches grow horizontally or may be somewhat inclined.

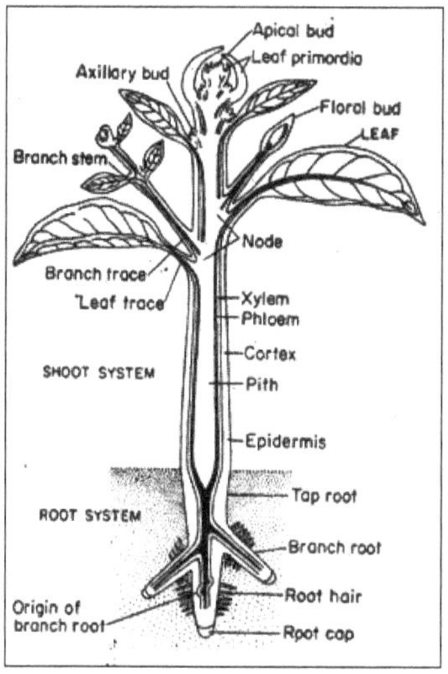

Ground plan of a plant.

On the stem are borne the similar branch stems, the leafy appendages that are dissimilar to the stem and the reproductive modifications of shoot which are commonly called flowers. The young stem is green but later on it may change its appearance because of secondary growth and the formation of bark. A stem may bear multicellular hairs and branches on it develop exogenously.

As opposed to the root, the nodes divide the stem into inter-nodes, the internodal Regions being bare while leaves and branches come out only from the nodal regions. If the nodes are very prominently developed, as in some grasses, the stem has a jointed or articulated appearance. A stem is usually cylindrical but it may be triangular (Cyperus rotundas), square (Labiatae), ribbed with ridges and furrows (Cucurbita), flat (Cactaceae), etc.

References

- Preparatory-events, plant-development, science: britannica.com, Retrieved 23 April, 2019
- Plants-growth-and-development-explained-with-diagram, plants, biology: yourarticlelibrary.com, Retrieved 25 May, 2019
- Stages-of-plant-growth: aeroponicsdiy.com, Retrieved 27 June, 2019
- Sporophyte, botany-general, botany, plants-and-animals: encyclopedia.com, Retrieved 29 July, 2019
- Sporophyte: biocyclopedia.com, Retrieved 2 August, 2019
- Bozhkov, P. V.; Filonova, L. H.; Suarez, M. F. (January 2005). "Programmed cell death in plant embryogenesis". Current Topics in Developmental Biology. 67: 135–179. doi:10.1016/S0070-2153(05)67004-4. ISBN 9780121531676. PMID 15949533
- Root-system, biology: byjus.com, Retrieved 4 January, 2019
- Plant-shoot-system-structure-function-quiz: study.com, Retrieved 6 February, 2019
- Shoot-system-of-plant-body-with-diagram, shoot-system: biologydiscussion.com, Retrieved 8 March, 2019

5

Animal Development

The processes that lead eventually to the formation of a new animal starting from cells derived from one or more parent individuals is known as animal development. Some of the stages of animal development are embryo formation, embryonic adaptations and organ formation. This chapter closely examines these key concepts as well as the stages of animal development to provide an extensive understanding of the subject.

Animal development is the processes that lead eventually to the formation of a new animal starting from cells derived from one or more parent individuals. Development thus occurs following the process by which a new generation of organisms is produced by the parent generation.

General Features

Reproduction and Development

In multicellular animals (Metazoa), reproduction takes one of two essentially different forms: sexual and asexual. In asexual reproduction the new individual is derived from a blastema, a group of cells from the parent body, sometimes, as in Hydra and other coelenterates, in the form of a "bud" on the body surface. In sponges and bryozoans, the cell groups from which new individuals develop are formed internally and may be surrounded by protective shells; these bodies, which may serve as resistant forms capable of withstanding unfavourable environmental conditions, are released after the death of the parent. In certain animals the parent may split in half, as in some worms, in which an individual worm breaks into two fairly equal parts (except that the anterior half receives the mouth, "brain," and sense organs if they are present).

Obviously, in such a case it is impossible to say which of the two resulting individuals is the parent and which the offspring. Some brittle stars (starfish relatives) may reproduce by breaking across the middle of the body disk, with each of the halves subsequently growing its missing half and the corresponding arms.

A common feature of all forms of asexual reproduction is that the cells—always a substantial number of cells, never only one cell—taking part in the formation of the new individual are not essentially different from other body, or somatic, cells. The number of chromosomes (bodies carrying the hereditary material) in the cells participating in the formation of a blastema is the same as in the other somatic cells of the parent, constituting a normal, double, or diploid (2n), set.

In sexual reproduction, a new individual is produced not by somatic cells of the parent but by sex cells, or gametes, which differ essentially from somatic cells in having undergone meiosis, a process in which the number of chromosomes is reduced to one-half of the diploid (2n) number found in somatic cells; cells containing one set of chromosomes are said to be haploid (n). The resulting sex cells thus receive only half the number of chromosomes present in the somatic cell. Furthermore, the sex cells are generally capable of developing into a new individual only after two have united in a process called fertilization.

Each type of reproduction—asexual and sexual—has advantages for the species. Asexual reproduction is, at least in some cases, the faster process, leading most rapidly to the development of large numbers of individuals. Males and females are independently capable of producing offspring. The large size of the original mass of living matter and its high degree of organization—the new individual inherits parts of the body of the parent: a part of the alimentary canal, for instance—make subsequent development more simple, and the attainment of a stage capable of self-support easier. New individuals produced by asexual reproduction have the same genetic constitution (genotype) as their parent and constitute what is called a clone. Though asexual reproduction is advantageous in that, if the parent animal is well adapted to its environment and the latter is stable, then all offspring will benefit, it is disadvantageous in that the fixed genotype not only makes any change in offspring impossible, should the environment change, but also prevents the acquisition of new characteristics, as part of an evolutionary process. Sexual reproduction, on the other hand, provides possibilities for variation among offspring and thus assists evolution by allowing new pairs of genes to combine in offspring. Since all body cells are derived from the fertilized egg cell, a mutation, or change, occurring in the sex cells of the parents immediately provides a new genotype in each cell of the offspring. In the course of evolution, sexual reproduction has been selected for, and established in, all main lines of organisms; asexual reproduction is found only in special cases and restricted groups of organisms.

Preparatory Events

In the case of multicellular animals we find there are two kinds of sex cells: the female sex cell (ovum, or egg), derived from an oocyte (immature egg), and the male sex cell (spermatozoon or sperm), derived from a spermatocyte. Eggs are produced in ovaries; sperm, in testes. Both the egg and the sperm contribute to the development of the new individual; each providing one set of genes, thereby restoring the diploid number of chromosomes in the fertilized egg. The sperm possesses a whiplike tail (flagellum) that enables it to swim to the egg to fertilize it. In most cases the egg, a stationary, spherical cell, provides the potential offspring with a store of food materials, or yolk, for its early development. The term yolk does not refer to any particular substance but in fact includes proteins, phosphoproteins, lipids, cholesterol, and fats, all of which substances occur in various proportions in the eggs of different animals. In addition to yolk, eggs accumulate other components and acquire the structure necessary for the development of the new individual. In particular the egg acquires polarity—that is, the two ends, or poles, of the egg become distinctive from each other. At one pole, known as the animal pole, the cytoplasm appears to be more active and contains the nucleus (meiotic divisions occur in this region); at the other, called the vegetal pole, the cytoplasm is less active and contains most of the yolk. The general organization of the future animal is closely related to the polarity of the egg.

When the amount of food reserve is comparatively small, as it is in many marine invertebrates and mammals (in the latter the embryo is nourished by materials in the mother's blood), the egg may be barely visible to the unaided eye. The egg of the sea urchin is about 75 microns (0.003 inch) in diameter; that of a human being is slightly more than 0.1 millimetre. Eggs are classified according to the amount of yolk present. An egg with a small quantity of evenly distributed yolk is called an oligolecithal egg. One with more yolk that is unevenly distributed (i.e., concentrated towards the vegetal pole) is telolecithal; and one with still greater amounts of yolk in granules or in a compact mass is megalecithal.

The egg is surrounded by protective membranes, which may be soft and jellylike or hard and calcified, like shells. Egg membranes are produced while the egg is either in the ovary or being carried away from the ovary in a tube called an oviduct. The eggs of many animals have both kinds of membranes. In insects, a hard shell (chorion) forms around the eggs in the ovaries. In frogs, a very thin vitelline membrane forms around the eggs in the ovary; subsequently a layer of jelly is deposited around the eggs while they pass through the oviducts. In birds, a very thin vitelline membrane is produced around the egg in the ovary; then several layers of secondary membranes are formed in the oviduct before the egg is laid. The outermost of these secondary membranes is the calcareous shell. In mammals the egg is surrounded by the so-called pellucid zone, which is equivalent to the vitelline membrane of other animals; follicle cells form an area called the corona radiata around this zone.

After fertilization the egg, now called a zygote, is endowed with genes from two parents and has begun actual development. (Activation of the egg may be brought about by an agent other than sperm in certain animals, but such cases of parthenogenesis are exceptional.

After fertilization, the zygote undergoes a series of transformations that bring it closer to the essential organization of the parents. These transformations, initiated at a physiological, perhaps even at a molecular, level, eventually result in the appearance of certain structures. The whole process is called. The process of development is more easily understood if, at every step, the changes necessary to bring the system nearer the goal are considered. Depending on the achievements necessary at any step, development can be subdivided into a number of discrete phases, the first of which, cleavage, immediately follows fertilization.

Embryo Formation

Embryonic Adaptations

Throughout its development the embryo requires a steady supply of nourishment and oxygen and a means for disposal of wastes. These needs are met in various ways, depending in particular on (1) whether the eggs develop externally (oviparity), are retained in the maternal body until ready to hatch (ovoviviparity), or are carried in the maternal body to a later stage (viviparity); and (2) the length of embryonic development.

Adaptations in Animals other than Mammals

Eggs of many marine invertebrates are discharged directly into water, and the period of development before the larva emerges is relatively brief. Oxygen diffuses easily into the small eggs, and nourishment is provided by a moderate amount of yolk. During cleavage the yolk is distributed to all the blastomeres. Much of the nourishment in the egg is stored as animal starch, or glycogen, which is almost completely used by the time the larva emerges from the egg. A small amount of water and inorganic salts are taken in by the embryo from surrounding seawater. Eggs developing in freshwater carry their own supply of necessary amounts of certain salts that are not present in sufficient quantities in the environment. Products of metabolism—especially carbon dioxide and nitrogenous wastes in the form of ammonia—diffuse out from small embryos developing in water.

The eggs of terrestrial animals must overcome the hazard of drying. In certain species this danger is avoided because the animal returns to water to breed, such as frogs and salamanders. Some groups of insects (e.g., dragonflies, mayflies, and mosquitoes) also lay eggs in water, and the larvae are aquatic. Eggs of other animals (e.g., snails, earthworms) are laid in moist earth and thus are protected from drying up. In terms of evolution, however, a decisive solution to the problem of development on land was arrived at by most insects and by reptiles and birds, which developed eggs with a shell impermeable to water or, at least, resistant to rapid evaporation. The shells of bird and insect eggs, while restricting evaporation of water, allow oxygen to diffuse into the egg and carbon dioxide to diffuse out. Apart from gas exchange, the eggs constitute closed systems, which give nothing to the outside and require nothing from it. Such eggs are called cleidoic. Because the products of nitrogen metabolism in cleidoic eggs cannot pass through the eggshell, animals (birds and insects) have had to evolve a method of storing wastes in the form of uric acid, which, since it is insoluble, is nontoxic to the embryo.

After a short period of development in the egg, the emerging young animal has to fend for itself, unless there is some form of parental care. Exposure to the external environment at a tender age results frequently in loss of life, a hazard met by many animals through an increase in the supply of nourishment within the egg, thus allowing the young to attain a greater size and development. This tendency to produce large yolky eggs has been achieved independently in different evolutionary lines: in octopuses and squids among the mollusks, in sharks among the fishes, and in reptiles and birds among the terrestrial vertebrates.

Cleavage is incomplete in eggs with large amounts of yolk. Although some yolk platelets may be enclosed in the formative cells of the embryo, the bulk of the yolk remains an uncleaved mass, overgrown and surrounded by the cellular part of the embryo. In such cases a membranous bag, or yolk sac, is formed and remains connected to the embryo by a narrow stalk (the evolutionary precursor of the umbilical cord of mammals). The cellular layers surrounding the yolk sac and forming its walls may consist of all three germinal layers (in reptiles and birds), so that the yolk virtually comes to lie inside an extension of the gut of the embryo; or (in bony fishes) the yolk sac may be enclosed in layers of ectoderm and mesoderm. In either case a network of blood vessels develops in the walls of the yolk sac and transports the yolk products to the embryo. As the yolk is broken down and utilized, the yolk sac shrinks and is eventually drawn into the body of the embryo. In addition to the yolk sac, extra-embryonic parts are also encountered in the form of embryonic membranes, which are found in higher vertebrates and in insects. Vertebrates have three embryonic membranes: the amnion, the chorion, and the allantois.

In reptiles, birds, and mammals, folds develop on the surface of the yolk sac just outside and around the body of the embryo proper. These folds, consisting of extra-embryonic ectoderm and extra-embryonic mesoderm, rise up and fuse dorsally, enclosing the embryo in a double-lined, fluid-filled chamber known as the amniotic cavity. The inner lining of the fold becomes the amnion, and the outer becomes the chorion, which ultimately surrounds the entire embryo. The amniotic fluid protects the embryo from drying, prevents the adhesion of the embryo to the inner surface of the shell, and provides the embryo with efficient shock absorption against possible damaging jolts. (The aminion and chorion develop in the same way in insect embryos.) The third membrane, or allantois, is originally nothing more than the urinary bladder of the embryo. It is a saclike growth of the floor of the gut, into which nitrogenous wastes of the embryo are voided. It enlarges greatly during the course of development, eventually expanding between the amnion and chorion and also between the chorion and the yolk sac, to become the third embryonic membrane. In addition to providing storage space for the nitrogenous wastes of the embryo, the allantois takes up oxygen, which penetrates into the egg from the exterior, and delivers it, by way of a network of blood vessels, to the embryo.

Adaptations in Mammals

At some early stage during the evolution of viviparous mammals, eggs came to be retained in the oviducts of the mother. The embryo then was provided with nourishment from fluids in the oviduct; the yolk, which became redundant, gradually ceased to be provided, and the eggs became oligolecithal. The eggshell, present in reptiles, was no longer needed and eventually disappeared, as did the white of the egg. The chorion, however, remained as the most external coat of the developing embryo through which nourishment reaches the embryo. It acquired the ability to adhere closely to the walls of the uterus (which was what that part of the oviduct holding the embryo had become) and became the so-called trophoblast. The blood-vessel network of the underlying allantois conveys nutrients that diffuse through the trophoblast to the body of the embryo proper. These modifications gave rise to a new organ, the placenta, formed from tissues of both the mother and the embryo: the uterine wall with its blood vessels provided by the mother; the trophoblast and allantois—and in some mammals also the yolk sac—with their blood vessels provided by the embryo.

The overall development of placental mammals as a result of these changes is profoundly different from that of their ancestors, the reptiles, and proceeds in the following way: the tiny yolkless egg is fertilized in the upper portion of the oviduct by sperm received from the male in the process of coupling (coitus); cleavage starts as the egg is propelled slowly down the oviduct by action of cilia in the oviduct lining. At the end of cleavage a solid ball of cells called a morula is produced. The surface cells of the morula become the trophoblast and the inner cell mass gives rise to the embryo (the formative cells) and also its yolk sac, amnion, and allantois. A cavity appears within the morula, converting it into a hollow embryo, called the blastocyst. This cavity resembles the blastocoel but, in fact, is analogous to the yolk sac of meroblastic eggs, except that there is no yolk and the cavity is filled with fluid. At the blastocyst stage, the embryo enters the uterus and attaches itself to the uterine wall. This attachment, or implantation, a crucial step in the development of a mammal, is attained through the action of the trophoblast, which forms extensions, known as villi that penetrate the uterine wall. In higher placental mammals, the lining of the uterine wall and, in varying degrees, the underlying tissues as well are partially destroyed, resulting in a closer relationship between the blood supplies of the mother and the embryo. Indeed, in man and in some rodents, the blastocyst sinks completely into the uterine wall and becomes surrounded by uterine tissue.

While implantation takes place, the formative cells arrange themselves in the form of a disk under the trophoblast. In the disk, the germinal layers develop much as in birds, with the formation of a primitive streak and migration of the chordamesoderm into a deeper layer. A layer of endoderm is formed adjoining the cavity of the blastocyst, and an amniotic cavity develops, enclosing the embryo; in lower placental mammals, the allantois also develops. The embryo proper, lying in the amniotic cavity, is connected to the extra-embryonic parts by the umbilical cord. The umbilical cord lengthens greatly during later development. In higher mammals, the cavity of the allantois is reduced, but the allantoic blood vessels become well developed and extend through the umbilical cord, connecting the embryo to the placenta. The blood that circulates in the placenta brings oxygen and nutrients from the maternal blood to the embryo and carries away carbon dioxide and other waste products from the embryo to the maternal blood for disposal by the maternal body.

Although tissues of maternal and embryonic origin are closely apposed in the placenta, there is little actual mingling of the tissues. Despite an occasional penetration of an embryo cell into the mother and vice versa, there is a placental barrier between the two tissues. The blood circulation of the mother is at all times completely separated from that of the embryo and its extra-embryonic parts. The placental barrier, however, does allow molecules of various substances to pass through; such differential permeability is indeed necessary if the embryo is to obtain nourishment. The permeability of the placental barrier differs in different animals; thus antibodies, which are protein molecules, may penetrate the placental barrier in man but not in cattle.

The maintenance of the fetus—as the more advanced embryo of a mammal is called—in the uterus is under hormonal control. In the initial stages of pregnancy, the continued existence of the embryo in the uterus depends on the hormone progesterone, which is secreted by the corpora lutea, "yellow bodies," that develop in the ovary after an egg has been released.

At birth the fetal parts of the placenta separate from the maternal parts. Contraction of the uterine wall first releases the fetus from the uterus; the fetal parts of the placenta (the afterbirth) follow. In certain cases of intimate connection between fetal and maternal tissues, the maternal tissues are torn, and birth is accompanied by profuse bleeding.

The Purpose of Animal Development

Body consist of over 30 trillion cells, but you began as a single cell: a fertilized egg, or zygote. How did you become the large, organized multicellular individual that you are today? Development.

As an animal embryo develops, its cells divide, grow, and migrate in specific patterns to make a more and more elaborate body (plant cells perform differential expansion instead of migration). To function correctly, that body needs well-defined axes (such as head vs. tail). It also needs a specific collection of many-celled organs and other structures, positioned in the right spots along the axes and connected up with one another in the right ways. How are all of these complexes processed accomplished and coordinated? They occur via four essential stages in early animal development:

- Fertilization: The process of a single sperm cell combining with single egg cell to form a zygote.
- Cleavage: Rapid, multiple rounds of mitotic cell division where the overall size of the embryo does not increase. The developing embryo is called a blastula following completion of cleavage.
- Gastrulation: The dramatic rearrangement (movement) of cells in the blastula to create the embryonic tissue layers. These tissue layers will go on to produce the tissues and organs of the adult animal.
- Organogenesis: The process of organ and tissue formation via cell division and differentiation.

Gastrulation and organogenesis together contribute to morphogenesis: the biological processes that result in an organism's shape and body organization.

Development Steps

Step 1: Fertilization

Fertilization is the process in which a single haploid sperm fuses with a single haploid egg fuse to form a zygote. The sperm and egg cells each possess specific features that make this process possible.

The egg is the largest cell produced in most animals species. Human eggs are approximately 16 times larger than a human sperm cell. The eggs of different species contain varying amounts of yolk, nutrients to support growth of the developing embryo. The egg is surrounded by a jelly layer, composed of glycoproteins (proteins that have sugars stuck to them), that releases species-specific chemoattractants (chemical-attractors) that activate and guide sperm to the egg. In mammals, this layer is called the zona pellucida. The zona pellucida/jelly layer is separated from the egg by a membrane called the vitalline envelope, which is outside of the cell's plasma membrane. Just underneath the egg's plasma membrane are cortical granules, vesicles containing enzymes that will degrade the proteins that hold the vitalline envelope around the plasma membrane when fertilization occurs.

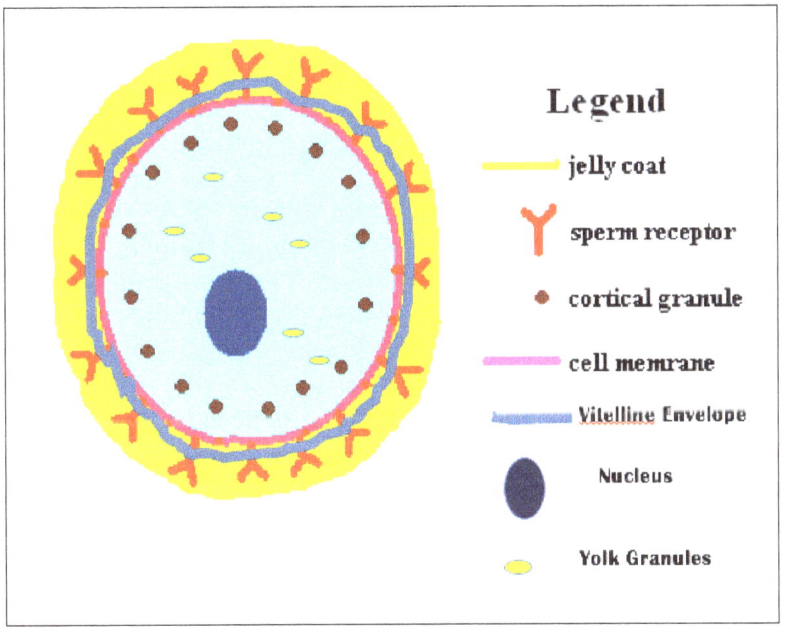

This diagram represents the components of a mammalian/human egg cell prior to the process of fertilization. It shows the outer jelly coating, the plasma (cell) membrane with sperm cell receptors and the cortical granules just inside the plasma (cell) membrane.

The sperm is one of the smallest cells produced in most animal species. The sperm consists of head containing tightly packed DNA, a flagellar tail for swimming, and many mitochondria to provide power for sperm movement. The plasma membrane of the sperm contains proteins called bindin, which are species-specific proteins that recognize and bind to receptors on the egg plasma membrane. In addition to the nucleus, the sperm head also contains an organelle called the acrosome, which contains digestive enzymes that will degrade the jelly layer/zona pellucida to allow the sperm to reach the egg plasma membrane.

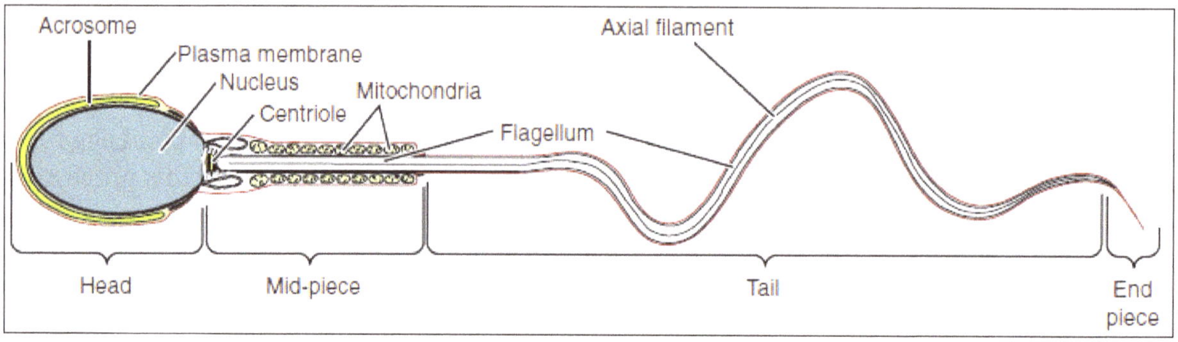

Sperm cells are divided into a head, containing DNA; a mid-piece, containing mitochondria; and a tail, providing motility. The acrosome is oval and somewhat flattened.

To ensure that the offspring has only one complete diploid set of chromosomes, only one sperm can fuse with one egg. Fusion of more than one sperm with an egg, or polyspermy, is genetically incompatible with life and results in zygote death. There are two mechanisms that prevent polyspermy: the "fast block" to polyspermy and the "slow block" polyspermy. These and the other steps of fertilization are described here:

1. Sperm is attracted to and contacts the jelly layer/zona pellucida.
2. Interactions receptors on the sperm cell and egg glycoproteins initiate the acrosome reaction. Digestive enzymes are released from the acrosome, and they destroy the jelly layer/zona pellucida thus creating a pathway for the sperm to approach the egg.
3. The sperm reaches the egg plasma membrane and the sperm and egg plasma membranes fuse. This interaction is mediated by the sperm bindin protein, which binds to bindin receptor proteins on the egg plasma membrane. This interaction is species-specific, with slightly different versions of the bindin protein and the bindin receptor proteins in every different species.
4. Sperm-egg membrane fusion initiates electrical depolarization of the entire egg plasma membrane, temporarily (for 10-20 seconds) preventing any other sperm from fusing with the egg plasma membrane. This membrane depolarization, mediated by an influx of sodium ions, is the fast block to polyspermy.
5. Membrane depolarization by sodium influx initiates a cross the wave of calcium released across the plasma membrane.
6. The calcium wave initiates the cortical reaction in the egg: the cortical granules fuse with the egg plasma membrane, releasing digestive enzymes that degrade the bindin receptor proteins that serve as sperm docking sites, and the proteins that hold the vitalline layer against the plasma membrane.
7. The cortical reaction results in lifting away of the vitalline layer away from the egg plasma membrane, to create the fertilization envelope. The fertilization envelope is a barrier that prevents additional sperm from reaching the egg, and is the slow block to polyspermy.
8. These events culminate in egg activation, causing the egg to recognize that fertilization has occurred and resulting in initiation of development.

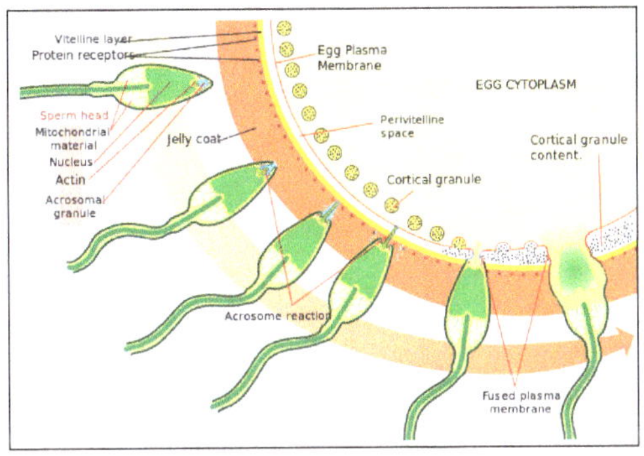

Fertilization is the process in which sperm and egg fuse to form a zygote. The acrosomal reaction help the sperm degrade the glycoprotein matrix protecting the egg and allow the sperm to transfer its nucleus.

Step 2: Cleavage and Blastula Stage

After fertilization successfully activates the egg, the egg begins a series of rapid cell divisions called cleavage, illustrated below. Normal cell division occurs every 18-24 hours, but cleavage cell divisions can occur as frequently as every 10 minutes. During cleavage, the cells divide without an increase in size; that is, one large single-celled zygote divides into multiple smaller cells called blastomeres. After the cleavage has produced over 100 blastomeres, the embryo is called a blastula. The blastula is usually a spherical layer of blastomeres that are considered to be the first embryonic tissue, the blastoderm. The blastoderm surrounds a fluid-filled or yolk-filled cavity, called the blastocoel (coelum = body cavity). The blastocoel is absolutely essential for the next step of development, gastrulation, which we will discuss in the next class session.

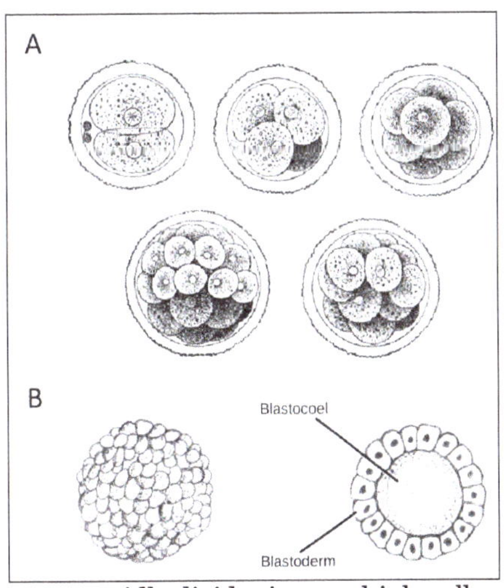

(a) During cleavage, the zygote rapidly divides into multiple cells without increasing in size.
(b) The cells rearrange themselves to form a hollow ball with a fluid-filled or yolk-filled cavity called the blastula.

The very early stages of mammalian development are extremely similar to other animals. Later stages of cleavage are a little different though, resulting in a structure called the blastocyst, which has an inner cell mass and an outer cell layer called the trophoblast. The inner cell mass will go on to form the embryo, and the trophoblast will contribute to the embryonic portion of placenta and nourish the embryo.

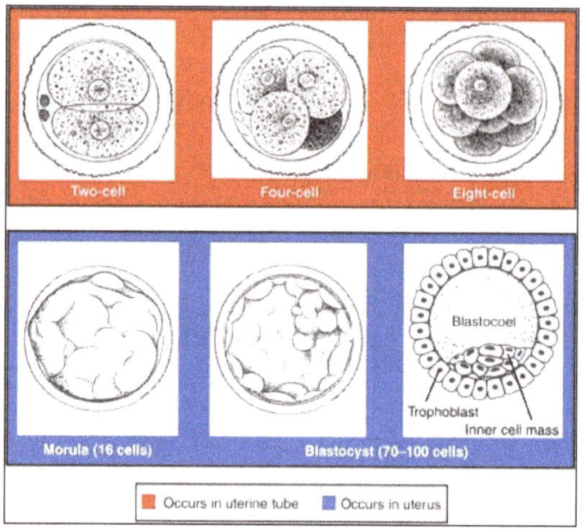

Pre-embryonic cleavages make use of the abundant cytoplasm of the fertilized egg as the cells rapidly divide without changing the total volume. The rearrangement of the cells in the mammalian blastula to two layers, the inner cell mass and the trophoblast, results in the formation of the blastocyst.

Intrinsic vs. Extrinsic Factors in Cell Specialization

Let's take a moment to discuss how cells "know" what they're supposed to do in an embryo, whether it's to divide, to migrate to a new location, or to begin specializing into a specific cell type.

Intrinsic (lineage) information is inherited from the mother cell, via cell division. For instance, a cell might inherit molecules that "tell" it that it belongs to the neural, or nerve cell-producing, lineage of the body.

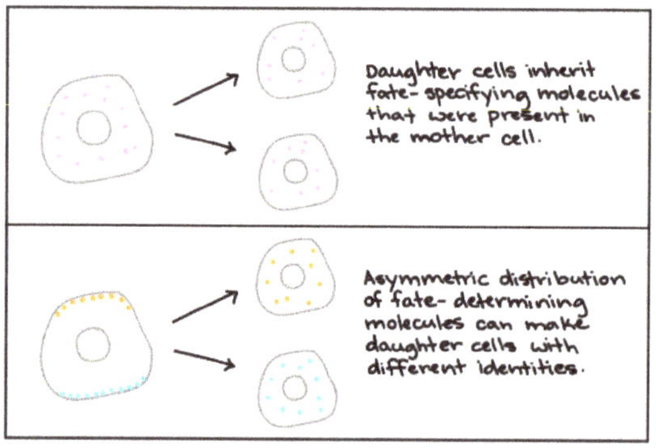

Intrinsic information is fate-specifying information found within a cell.

Extrinsic (positional) information is received from the cell's surroundings. For instance, a cell might get chemical signals from a neighbor, instructing it to become a particular kind of photoreceptor (light-detecting neuron).

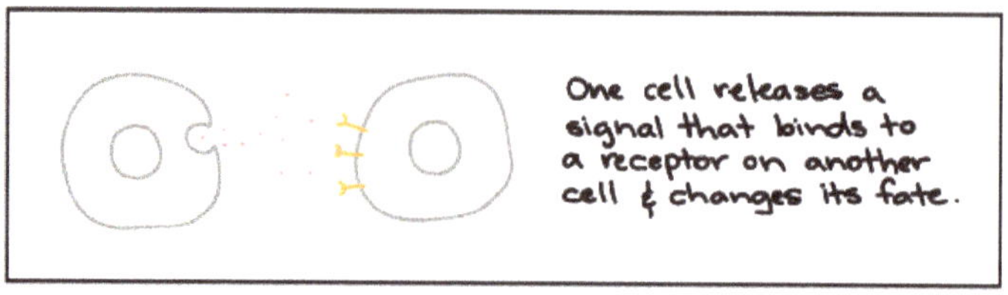

Extrinsic information is fate-specifying information received from outside of the cell.

Determining the Body Axes: Cytoplasmic Determinants and Induction

One of the fundamental principles of animal development in all animals (except sponges) is the establishment of the body axes: animal bodies have lateral-medial (left-right), dorsal-ventral (back-belly), and anterior-posterior (head-feet) axes.

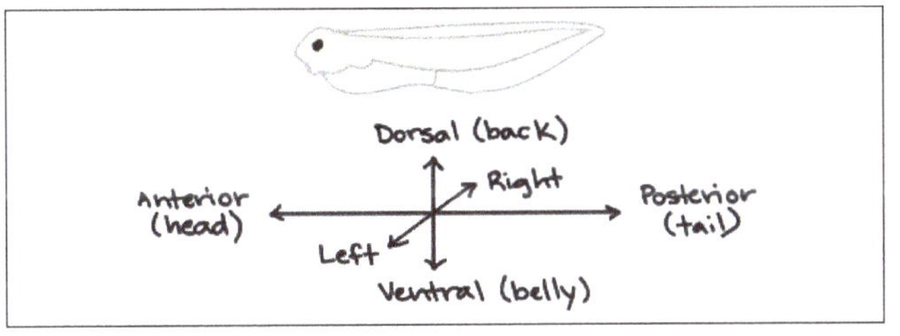

Animals have three body axes: anterior-posterior, dorsal-ventral, and left-right.

How are these axes established from a ball of apparently identical cells (the blastula)? The process is different among different lineages of animals, which body axes being heavily influenced by cytoplasmic determinants in protostomes (most invertebrates), by yolk polarity in vertebrates with large amounts of asymmetrically-distributed yolk (many fish, amphibians, reptiles, and birds), and by induction (cell-cell communication) in many mammals:

1. Cytoplasmic determinants are mRNAs or proteins found in the egg prior to fertilization (they come from mom's genome, not the embryo). They are asymmetrically distributed, so that after the first cleavage division, the two different resulting cells have different intrinsic information, which will then lead to different cell fates. One of the best understood cytoplasmic determinants is a factor called bicoid, which is present in a concentration gradient across the unfertilized eggs of Drosophila (fruit flies). The region of the egg with the highest concentration of bicoid becomes the anterior (head) portion of the embryo, while the region with the lowest bicoid concentration becomes the posterior (tail) region of the embryo. This future cell identity of anterior vs posterior is set after the very first cleavage division, where one cell gets nearly all the bicoid and the other cell has almost none. Cytoplasmic determinants are a key feature of protostome development and some deuterostomes, but they are not present in mammalian embryos.

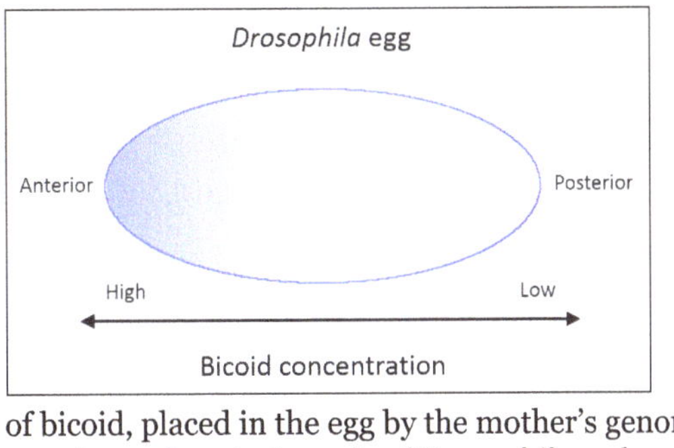

The distribution of bicoid, placed in the egg by the mother's genome, determines the anterior and posterior axis of Drosophila embryos.

2. Yolk polarity is a phenomenon that occurs in the eggs of animals which have large amounts of yolk in their eggs, such as amphibians, reptiles, birds, and many fish. Because the egg contains so much yolk, and the yolk is literally heavier than the rest of the egg, it sinks to

the bottom of the egg due to gravity. The yolk is therefore asymmetrically distributed in the egg, and the region of the egg with very little yolk will go on to become the anterior (head) structures of the embryo, while the region of the egg containing the mass of yolk will go on to become the posterior (tail) structures of the embryo.

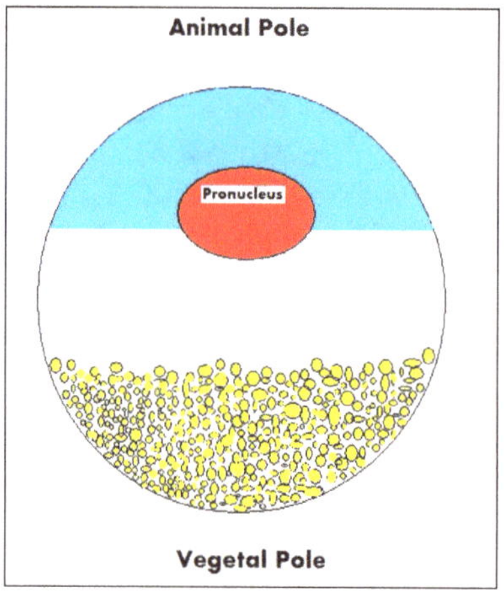

The cells arising from the animal pole of amphibian eggs will go on to become the anterior (head) region of the embryo, while the cells arising from the vegetal pole will go on to become the posterior (tail) region of the embryo.Cells neare the top.

3. Induction, or cell-cell communication, is one of the five essential developmental processes, and occurs during development of all animal (and plant) species. Induction is a prime example of extrinsic information that leads to different cell fates among initially identical cells. Because mammalian embryos have no cytoplasmic determinants and have very small amounts of evenly-distributed yolk, induction is the primary process responsible for establishing the body axes in mammalian embryos.

Stages of Animal Development

Throughout the animal kingdom, an incredible variety of embryonic types exist, but most patterns of embryogenesis are variations on five themes:

1. Immediately following fertilization, cleavage occurs. Cleavage is a series of extremely rapid mitotic divisions wherein the enormous volume of zygote cytoplasm is divided into numerous smaller cells. These cells are called blastomeres, and by the end of cleavage, they generally form a sphere known as a blastula.

2. After the rate of mitotic division has slowed down, the blastomeres undergo dramatic movements wherein they change their positions relative to one another. This series of extensive cell rearrangements is called gastrulation, and the embryo is said to be in the gastrula stage. As a result of gastrulation, the embryo contains three germ layers: the ectoderm, the endoderm, and the mesoderm.

3. Once the three germ layers are established, the cells interact with one another and rearrange themselves to produce tissues and organs. This process is called organogenesis. Many

organs contain cells from more than one germ layer, and it is not unusual for the outside of an organ to be derived from one layer and the inside from another. For example, the outer layer of skin comes from the ectoderm, while the inner layer (the dermis) comes from the mesoderm. Also during organogenesis, certain cells undergo long migrations from their place of origin to their final location. These migrating cells include the precursors of blood cells, lymph cells, pigment cells, and gametes. Most of the bones of our face are derived from cells that have migrated ventrally from the dorsal region of the head.

4. In many species a specialized portion of egg cytoplasm gives rise to cells that are the precursors of the gametes (the sperm and egg). The gametes and their precursor cells are collectively called germ cells, and they are set aside for reproductive function. All the other cells of the body are called somatic cells. This separation of somatic cells (which give rise to the individual body) and germ cells (which contribute to the formation of a new generation) is often one of the first differentiations to occur during animal development. The germ cells eventually migrate to the gonads, where they differentiate into gametes. The development of gametes, called gametogenesis, is usually not completed until the organism has become physically mature. At maturity, the gametes may be released and participate in fertilization to begin a new embryo. The adult organism eventually undergoes senescence and dies.

5. In many species, the organism that hatches from the egg or is born into the world is not sexually mature. Indeed, in most animals, the young organism is a larva that may look significantly different from the adult. Larvae often constitute the stage of life that is used for feeding or dispersal. In many species, the larval stage is the one that lasts the longest, and the adult is a brief stage solely for reproduction. In the silkworm moths, for instance, the adults do not have mouthparts and cannot feed. The larvae must eat enough for the adult to survive and mate. Indeed, most female moths mate as soon as they eclose from their pupa, and they fly only once—to lay their eggs. Then they die.

Ectodermal Derivatives

The Nervous System

The vertebrate nervous system develops from the neural plate—a thickened dorsal portion of the ectoderm—which forms a tube, as described earlier. From the very start the tube is wider anteriorly, the end that gives rise to the brain. The posterior part of the neural tube, which gives rise to the spinal cord, is narrower and stretches as the embryo lengthens. Stretching involves the head to only a very minor degree.

The Brain and Spinal Cord

Constrictions soon appear in the brain region of the neural tube, subdividing it into three parts, or brain vesicles, which undergo further transformations in the course of development. The most anterior of the primary brain vesicles, called the prosencephalon, gives rise to parts of the brain and the eye rudiments. The latter appear in a very early stage of development as lateral protrusions from the wall of the neural tube, which are constricted off from the remainder of the brain rudiment as the optic vesicles. The rest of the prosencephalon constricts further into two portions, an anterior one, or telencephalon, and a posterior one, or diencephalon. The telencephalon gives rise, in lower vertebrates, to the smell, or olfactory, centre; in higher vertebrates and man, it becomes the centre of mental activities. The diencephalon, with which the eye vesicles are connected, was presumably originally an optic centre, but it has acquired, in the course of evolution, a function of hormonal regulation. The floor of the diencephalon forms a funnel-shaped depression, the infundibulum, which becomes connected with the pituitary, or hypophysis, the most important gland of internal secretion (i.e., endocrine gland) in vertebrates. Indeed, the posterior lobe of the hypophysis is actually derived from the floor of the diencephalon. Tissues of the infundibulum and the posterior lobe of the hypophysis produce certain hormones (oxytocin and vasopressin) and stimulate the production and release of other hormones from the anterior lobe of the hypophysis.

The second primary brain vesicle, the mesencephalon, gives rise to the midbrain, which, in higher vertebrates, takes part in coordinating visual and auditory stimuli.

The third primary brain vesicle, the rhombencephalon, is more elongated than the first two; it produces the metencephalon, which gives rise to the cerebellum with its hemispheres, and the myelencephalon, which becomes the medulla oblongata. The cerebellum acts as a balance and coordinating centre, and the medulla controls functions such as respiratory movements.

The cells constituting the wall of the neural tube and, later, of the brain and spinal cord become arranged in such a way that they point into the central cavity of the tube. The differentiation of nervous tissue involves many cells abandoning their connection to the inner surface of the neural tube and migrating outward, where they accumulate as a mantle. The first cells to migrate become the neurons, or nerve cells. They produce outgrowths called axons and dendrites, by which the cells of the nervous system establish communication with one another to form a functional network. Some of the outgrowths extend beyond the confines of the brain and spinal cord as components of nerves; they establish contact with peripheral organs, which thus fall under the control of the nervous system. Cells migrating from the inner surface of the neural tube later in development become astrocytes, which are the supporting elements of nerve tissue.

The fate of nerve cells is dependent largely on whether they succeed, directly or indirectly (through other neurons), in connecting with peripheral organs. Nerve cells that fail to establish connections die. Thus, if in early stages of embryonic development, some organ, a limb rudiment for instance, is surgically removed, the nerve cells in the centres supplying nerves to such an organ are reduced in number, and the corresponding nerves also diminish or disappear. On the other hand, if an organ is introduced by transplantation into a developing embryo, the organ will be supplied by nerves from a nerve centre in which the number of cells apparently increases; no additional cells are provided, but cells that would otherwise have degenerated remain active and differentiate into functional neurons, thus satisfying the demand created by the additional organ.

Nerves do not consist entirely of outgrowths of neurons located in the brain and spinal cord. Many components of nerves are outgrowths of neurons, the cell bodies of which are located in masses called ganglia; there are three main types of ganglia: spinal ganglia, cranial ganglia, and ganglia of the autonomous nervous system. The spinal ganglia are derived from cells of the neural crest—the loose mesenchyme-like tissue that remains between the neural tube and skin after separation of the two. Part of the cells of the neural crest in the region of the trunk and tail accumulate in segmental groups (corresponding to the mesodermal somites) and provide fibres to peripheral organs and to the spinal cord. These fibres constitute the sensory pathways in the spinal nerves. The motor components of the spinal nerves—fibres that activate muscles—are outgrowths of neurons lying in the spinal cord. The ganglia of the cranial nerves are produced only in part from cells of the neural crest; an additional component comes from the epidermis on the side of the head. Cells of the epidermal thickenings called placodes detach themselves and contribute to the formation of the cranial ganglia and thus of the cranial nerves.

The ganglia of the autonomous (sympathetic) nervous system are derived, as are the spinal ganglia, from neural-crest cells, but, in this case, the cells migrate downward to form groups near the dorsal aorta, near the intestine, and even in the intestinal wall itself. The outgrowths of cells in these ganglia are the nerve fibres of the sympathetic nerves.

Major Sense Organs

The Eye

As has been pointed out, the rudiments of the eyes develop from optic vesicles, each of which remains connected to the brain by an eye stalk, which later serves as the pathway for the optic nerve. The optic vesicles extend laterally until they reach the skin, whereupon the outer surface caves in so that the vesicle becomes a double-walled optic cup. The thick inner layer of the optic cup gives rise to the sensory retina of the eye; the thinner outer layer becomes the pigment coat of the retina. The opening of the optic cup, wide at first, gradually becomes constricted to form the pupil, and the edges of the cup surrounding the pupil differentiate as the iris. The refractive system of the eye and, in particular, the lens of the eye are derived not from the cup but from the epidermis overlying the eye rudiment. When the optic vesicle touches the epidermis and caves in to produce the optic cup, the epidermis opposite the opening thickens and produces a spherical lens rudiment. The lens develops by an induction by the optic vesicle on the epidermis with which it comes in contact. A further influence emanating from the eye changes the epidermis remaining in place over the lens into a transparent area, the cornea. Influence of the optic cup on the surrounding mesenchyme causes the latter to produce a vascular layer around the retina and, outside of that, a tough fibrous or (in some animals) even a partly bony capsule called the sclera. Thus a complex interdependence of different materials produces the fully developed and functional vertebrate eye.

The Ear

The main part of the ear rudiment is derived from thickened epidermis adjoining the medulla. This area of the epidermis invaginates to produce the ear vesicle, which separates from the epidermis but remains closely opposed to the medulla. The ear vesicle becomes complexly folded to produce the labyrinth of the ear. Subsequently, a group of cells of the ear vesicle becomes detached and gives rise to the acoustic ganglion. Neurons of this ganglion become connected by their nerve fibres to the sensory cells in the labyrinth, on the one hand, and with the brain (the medulla), on the other. The ear vesicle, acting on the surrounding mesenchyme, induces the latter to aggregate around the labyrinth and form the ear capsule. Further parts with various origins are added to the ear: the middle ear, from a pharyngeal pouch and the associated skeleton, and the external ear (where present), from epidermis and dermis.

The Olfactory Organ

The olfactory organ develops from a thickening of the epidermis adjacent to the neural fold at the anterior end of the neural plate. This thickening is converted into a pocket or sac but does not lose connection with the exterior. The openings of the sac become the external nares, and the cavity of the sac becomes the nasal cavity. Some cells of the olfactory sac differentiate as sensory epithelium and produce nerve fibres entering the forebrain. In most fishes the olfactory sac does not communicate with the oral cavity; in lungfishes and in terrestrial vertebrates, however, canals develop from the olfactory sacs to the oral cavity, where they open by internal nares. A cartilaginous capsule forms around the olfactory organ from cells believed to have been derived from the walls of the sac itself, and thus it is ectodermal in origin.

Gustatory and other Organs

Gustatory organs in the form of taste buds develop as local differentiations of the lining of the oral cavity but also, in fishes, in the skin epidermis. They are supplied with nerve endings, as are several other sensory bodies scattered among the tissues and organs of the developing body.

The Epidermis and its Outgrowths

The major part of the ectodermal epithelium covering the body gives rise to the epidermis of the skin. In fishes and aquatic larvae of amphibians, the many-layered epidermis is provided with unicellular mucous glands. In terrestrial vertebrates, however, the epidermis becomes keratinized; i.e., the outer layers of cells produce keratin, a protein that is hardened and is impermeable to water. During the process of keratinization, many cell components degenerate and the cells die; the layer of keratinized cells is therefore shed from time to time. In reptiles the shedding may take the form of a molt in which the animal literally crawls out of its own skin. It is less well known that frogs and toads also molt, shedding the surface keratinized layer of their skin (which is usually eaten by the animal). In birds and mammals, keratinized cells are shed in pieces that are sloughed off, rather than in extensive layers. In many vertebrates local thickenings of the keratinized layer appear in the form of claws, hooves, nails, and horns.

The epidermis is only the superficial layer of the skin, which is reinforced by the dermis, a connective tissue layer of a much greater thickness. The cells of the dermis are derived from mesoderm and neural-crest cells. In particular the pigment cells found in the dermis of fishes, amphibians, and reptiles are of neural-crest origin. The pigment in the skin of birds and mammals (and also in hairs and feathers) is also produced by neural-crest cells, but in these animals the pigment cells penetrate into the epidermis or deposit their pigment granules there.

The structure of the skin is further complicated by the development of hairs and feathers, on the one hand, and of skin glands, on the other. Hairs and feathers develop from a somewhat similar kind of rudiment. The development starts with a local thickening of the epidermal layer, beneath which a group of mesenchyme cells accumulate. In the case of hairs, the epidermal thickening proliferates downward and forms the root of the hair, from which the shaft then grows outward, emerging on the surface of the skin. In the case of feathers, the epidermal thickening bulges outward to form a hollow fingerlike protrusion with a connective tissue core. Secondarily, the shaft of the feather branches characteristically to produce barbs and barbules. In both cases, however, the final structure—shaft of the hair and shaft barbs and barbules of the feather—consists of keratinized and, thus, dead cells.

The skin of amphibians and mammals (but not of birds and reptiles) is provided with numerous skin glands, which develop as ingrowths from the epidermis. A peculiar type of skin gland is the mammary gland of placental mammals. In the first stage of development, mammary-gland rudiments resemble hair rudiments; they are thickenings of the epidermis, with condensed mesenchyme on their inner surfaces. In some mammals (rabbit, man) two continuous epidermal thickenings called mammary lines stretch along either side of the belly of the embryo. Parts of the line corresponding in number and position to the future glands enlarge while the rest of the thickening disappears. The initial thickenings proliferate inward and produce a system of ramified cords, solid at first but hollowed out later, which become the lactiferous, or milk-bearing, ducts of the gland. Further branching at the tips of the ducts gives rise to smaller ducts and to the secretory end sacs, or alveoli, of the gland.

Mesodermal Derivatives

The Body Muscles and Axial Skeleton

The somites, formed in the early stages of development from the upper edges of the mesodermal mantle adjoining the notochord, are complex rudiments that subdivide and give rise to very diverse body structures. The coelomic cavity, present initially, becomes obliterated by the side-to-side flattening of the somites, so that the thinner, outer parietal layer of the somite comes in close contact with its thicker visceral layer. The visceral layer of the somite very early subdivides into two parts. The upper, dorsolateral part called the myotome remains compact, giving rise to the body muscles. The lower, medioventral part of the somite, called the sclerotome, breaks up into mesenchyme, which contributes to the axial skeleton of the embryo—that is, the vertebral column, ribs, and much of the skull. The parietal layer of the somite, at a later stage, is converted into mesenchyme that, together with components of the neural crest, gives rise to the dermis of the skin and, for this reason, is called the dermatome.

The cells of the myotome are elongated in a longitudinal direction and become differentiated as muscle fibres. The myotomes, originally situated dorsally, expand on either side, penetrating between the skin on the outside and the lateral plates of the mesoderm on the inside, until they meet midventrally; the whole body is thus enclosed in a layer of developing muscle. As the somites and myotomes are segmented, so are the muscles derived from them. Metamerism, or segmentation, a feature in the embryos of all vertebrates, remains preserved only in the adults of fishes and of terrestrial vertebrates that have elongated bodies (salamanders, snakes); it becomes largely erased in four-footed animals that depend on their limbs for locomotion.

The mesenchyme derived from the sclerotomes condenses as cartilage around the notochord and the spinal cord. It forms the cartilaginous vertebral column and ribs. In the head region it produces a part of the cartilaginous skull, mainly its posterior and ventral parts; anteriorly the somitic mesenchyme is supplemented by mesenchyme from the neural crest. Cartilaginous capsules of the olfactory organ and the ear fuse with the cartilaginous capsule surrounding the brain; to this complex are also added cartilages associated with the jaws and gill skeleton. Cartilage in the vertebral column and in the skull is replaced later in the bony fishes and in the terrestrial vertebrates by bone. At a still later stage, dermal bones are added, which, while they have no precursors in the cartilaginous skeleton, develop in the adjoining mesenchyme.

The Appendages: Tail and Limbs

The tail in vertebrates is a prolongation of the body beyond the anus. It develops in early stages from the tail bud, immediately dorsal to the blastopore. Material for the tip of the tail is situated slightly forward from the edge of the blastopore. The elongation of the back of the body is greater than that of the belly; as a result the tip of the tail bud is carried beyond the blastopore and thus beyond the anus, which, in the developed embryo, marks the position of the blastopore. The consequence is that a section of the dorsal surface of the embryo comes to lie on the ventral surface of the tail; i.e., becomes inflected. The tail bud is formed from parts that have already been differentiated to a certain extent; prolongations of the neural tube and of the notochord are involved, and endoderm extends into the tail rudiment as the postanal gut, which, however, soon degenerates. The bud is also encased in ectodermal epidermis. In amphibians the somites of the tail are not derived from the chordamesodermal mantle but from the inflected posterior portion of the neural plate, which loses its nervous nature and becomes subdivided into segments corresponding to the somites of the trunk. In higher vertebrates the cells in the interior of the tail bud have an undifferentiated appearance and form a growth zone, at the expense of which parts of the tail (neural tube, notochord, somites) are extended backward as the tail elongates.

The paired limbs of vertebrates derive their first rudiments from the upper edge of the lateral plate mesoderm. The parietal layer becomes thickened, and cells escape from the epithelial arrangement and form a mesenchymal mass adjoining the ectodermal epithelium at the surface of the body. The ectodermal epithelium over the mass of mesenchyme likewise becomes thickened. In higher vertebrates, the accumulation of mesodermal cells and the thickening of the epidermis occur along the entire length of the trunk, from neck to anus, but in the middle of the trunk they soon disappear, and only the most anterior and the most posterior sections develop further into the rudiments of the forelimbs and hindlimbs, respectively. In fishes, the rudiments of the pectoral and pelvic fins are more extended anteroposteriorly in earlier than in final stages.

The mesodermal masses of the limb rudiments proliferate, and, covered with thickened epidermis, form on the surface of the body conical protrusions called the limb buds, which, once formed, possess all the materials necessary for limb development. Limb buds may be transplanted into various positions on the body or on the head and there develop into clearly recognizable limbs, conforming to their origin, whether a forelimb or hindlimb, a wing or a leg in birds. This specificity of the limb is carried by the mesodermal part of the rudiment, but a complex interaction between the mesodermal mesenchyme and the ectodermal epidermis is necessary for the normal development of the limb. In four-limbed vertebrates (tetrapods), the tips of the limb buds become flattened and broadened into hand or foot plates. The edge of the plate is indented, forming the rudiments of the digits. Meanwhile, local areas of the mesodermal mesenchyme in the interior of the limb rudiment condense; these are the rudiments of the various components of the limb skeleton. In fishes, small outgrowths from the myotomes enter the limb rudiment to form the muscles of the fins. In tetrapods, however, the limb muscles develop from the same mass of mesenchyme that gives rise to the skeleton. Thus the muscles of the body and the muscles of the limbs have different origins—the first develop from the myotomes (thus from the somites), and the second develop from the lateral plate mesoderm via the limb buds.

The nerves supplying the limbs grow into the limb rudiments from the spinal cord and the spinal ganglia. The nerves are guided in some way by the limb rudiments, for, if limb rudiments are displaced by transplantation to an abnormal position, the nerves still find their way and establish normal relationships to the limb muscles. Limb rudiments transplanted to sites very far from their normal positions induce local nerves to enter the limb, thereby making it motile.

Excretory Organs

The kidneys of vertebrates consist of a mass of tubules that develop from the stalks of somites called nephrotomes. In some primitive vertebrates such as cyclostomes, the nephrotome in each segment gives rise to only one tubule, but, in the great majority of vertebrates, mesenchyme from adjacent nephrotomes fuses into a common mass that differentiates into a number of nephric tubules irrespective of the original segmentation of the mesoderm. Under primitive conditions each tubule opens by a funnel (the nephrostome) into the coelomic cavity; the opposite ends of the tubules fuse to form the collecting ducts of the kidney. A collection of capillaries (the glomerulus) becomes associated with the nephric tubule, forming its filtration apparatus. The glomerulus may be situated in the coelomic cavity opposite the nephrostome or, in all the more advanced animals, intercalated into the nephric tubule, forming with the latter a renal corpuscle of the kidney. In adults of all vertebrates above the amphibians, the nephrostomes disappear (or are never formed), so that the tubule begins with the renal corpuscle. Parts of the kidney in vertebrates can be distinguished as the pronephros (most anteriorly, at the forelimb level), the mesonephros (in the midtrunk region), and the metanephros (in the pelvic region). The three sections of the kidney develop at different stages, starting with the pronephros and ending with the metanephros. In their morphology and mode of development, the anterior parts show more primitive conditions than the posterior ones. The pronephros, developing early in embryo formation, is the functional kidney of fish and amphibian larvae. Its collecting duct opens into the hindmost part of the intestine, called the cloaca, and later also serves as the collecting duct of the mesonephros. In reptiles, birds, and mammals, the pronephros is nonfunctional, although even in these animals its duct persists as the mesonephric duct. The mesonephros develops later and replaces the pronephros as the functional kidney of adult fishes and amphibians and of the embryos of reptiles, birds, and mammals. The tubules of the mesonephros link up with the duct derived from the pronephros. The pronephric duct in fact stimulates the development of mesonephric tubules, and, in its absence, the mesonephros does not develop at all.

The metanephros is found only in reptiles, birds, and mammals. It replaces the mesonephros of the early embryonic stages and continues as the functional kidney in the postembryonic and adult life of these animals. The metanephros develops from mesenchyme derived from the nephrotomes of the posterior part of the trunk and lying dorsal to the mesonephric duct. The actual differentiation is initiated by a dorsal outgrowth of the mesonephric duct, called the ureteric bud. The ureteric bud grows in the direction of the mesenchyme and becomes the ureter. Having penetrated the mass of mesenchyme, it starts to branch, producing the collecting tubules of the kidney; the mesenchyme, meanwhile, in response to the influence of the duct and its branches, aggregates to form the excretory tubules of the kidney. The influence of the ureter is indispensable for the development of the metanephric excretory tubules, for, if the ureter fails to develop or, in its outgrowth, stops short of reaching the kidney-producing mesenchyme, no kidney develops.

Circulatory Organs

The rudiment of the heart in vertebrates develops from the ventral edges of the mesodermal mantle in the anterior part of the body, immediately adjoining the pharyngeal region. A group of mesodermal cells breaks away from the ventral edge of the lateral plate, takes a position just underneath the pharyngeal endoderm, and becomes arranged in the form of a thin-walled tube, which will become the endocardium, or lining of the heart. In vertebrates with complete cleavage, the endocardial tube is single and medial from its start. In higher vertebrates with meroblastic cleavage—reptiles, birds, and mammals—the embryo in early stages of development is flattened out on the surface of the yolk sac; therefore, what are morphologically the ventral edges of the mesodermal mantle lie far apart on the perimeter of the blastodisc. As a result of this arrangement, two endocardial tubes are formed, one on either side of the embryo. Subsequently, when the embryo becomes separated from the yolk sac, the two endocardial tubes meet in the midline ventral to the pharynx and fuse, producing a single heart rudiment. After the formation of the endocardium, or the lining of the heart, the coelomic cavity in the lateral plate mesoderm adjoining the heart rudiment expands slightly and envelops the endocardial tube or tubes. The heart muscle layer, or myocardium, develops from the visceral (splanchnic) layer of the lateral plate that is in contact with the endocardial tube; the parietal (somatic) layer of the lateral plate forms the pericardium, or covering of the heart. The portion of the coelom surrounding the heart becomes separated from the rest of the body cavity and develops into the pericardial cavity.

The endocardial tube branches anteriorly into two tubes, the ventral aortas; a similar branching of the endocardial tube posteriorly forms the two vitelline veins, which carry blood from the midgut endoderm or from the yolk sac (when present) to the heart.

In its earliest development, the heart rudiment shows a degree of dependence on the adjoining endoderm. The whole of the endoderm can be removed in newt embryos in the neural-tube stage. In such endodermless embryos, the heart fails to develop, even though the mesoderm destined to form the heart rudiment is left intact.

The heart is initially a straight tube stretching in an anteroposterior direction. Rather early in development, however, it becomes twisted in a characteristic way and subdivided into four main parts: the most posterior, the sinus venosus; the atrium, which comes to lie at the anteriorly directed bend of the tube; the ventricle, occupying the apex of the posteroventrally directed inflexion; and, most anteriorly, the conus arteriosus. In the course of development in the more advanced vertebrates, the atrium and ventricle become partially or completely subdivided into right and left halves. In amphibians, only the atrium is separated into two halves, by a partition starting from the posterior end. In reptiles, a partition separates the atria and part of the ventricle. In birds and mammals, the subdivision of the heart is complete, with two atria and two ventricles.

The complete subdivision of the heart is important for separating the pulmonary, or lung, blood supply from the general body circulation. But, if this separation developed early in the embryo, it would create difficulties, since the lungs of the embryo are not functional; the enrichment of the blood with oxygen occurs instead in the placenta. The partition between the atria in mammalian embryos remains incomplete, so that blood returning from the body and from the placenta enters into the right half of the heart but is shunted (through the interatrial foramen) into the left half of the heart and thence again into general circulation. At birth, however, the interatrial foramen is closed by a membraneous flap, and oxygen-depleted blood from the body enters the right atrium, is channelled into the right ventricle, and thence to the lungs for oxygenating.

In an adult vertebrate, blood vessels extend to all parts of the body. It would seem that channels for the supply of blood are provided in proportion to the local demand of the tissues; progressively developing organs or parts with particularly intensified function always receive an increased blood supply. The rudiments of blood vessels are always aggregations of mesenchyme cells. In any blood vessel the endothelial tube is formed first, and the muscular and elastic layers are added later.

The main blood channels in vertebrates develop in certain favoured situations; namely (1) between the endoderm and lateral plate mesoderm; (2) around the kidneys, especially the pronephros and mesonephros; and (3) in connection with the heart, which is a special case of the first category.

From the paired forward extensions from the heart, the ventral aortas, loops develop between the pharyngeal clefts. These are the aortic arches, which served originally to supply blood to the gills in aquatic vertebrates. The arches are laid down in all vertebrates, six or more being found in cyclostomes and fishes; six are present in the embryos of tetrapods, but the first two are degenerate. The arches of the third pair develop as the carotid arteries, supplying blood to the head. Those of the fourth pair (and, exceptionally, in urodeles also the fifth) join dorsally to form the dorsal aorta, providing blood to most of the body. These are the systemic arches. The arches of the sixth pair are the pulmonary arches; in embryos they carry blood to the dorsal aorta, as well as to the lungs, but in fully developed amniotes (reptiles, birds, and mammals), they carry blood only to the lungs.

The paired posterior extensions of the heart of the early embryo are the vitelline veins, whose branches spread out between the lateral plate mesoderm and the endoderm, especially the endoderm of the yolk sac, when present. On their way to the heart, the vitelline veins pass through the liver and break up into a system of small channels—the hepatic sinusoids. Parts of the vitelline veins lying posterior to the liver become the hepatic portal veins, which carry blood from the intestine to the liver; the parts of the vitelline veins anterior to the liver become the hepatic veins, which carry blood from the liver to the sinus venosus in lower vertebrates (anamniotes), but become the anterior section of the postcaval vein in amniotes.

Whereas the vitelline veins and, later, the hepatic portal vein carry blood from the endodermal parts of the embryo and from the yolk sac to the heart, the blood from the mesodermal and ectodermal parts is returned to the heart through a system of cardinal veins. These latter veins start their development in the form of an irregular sinus around the pronephros, connected by the common cardinal veins (ducts of Cuvier), on either side, to the sinus venosus. Extensions anteriorly and posteriorly give rise to the precardinal and postcardinal veins, respectively. The postcaval vein, present in terrestrial vertebrates, is a late acquisition, both in evolution and in embryogenesis; it is a result of the intercommunication of several venous channels, including the anterior portion of the vitelline veins.

The first blood cells in vertebrate embryos form in association with the intestinal endoderm on the yolk sac. Groups of mesoderm cells derived from the splanchnic layer of the lateral plate (extra-embryonically in cases in which a yolk sac is present) become so-called blood islands, which are particularly conspicuous on the yolk sac of bird embryos (in the area vasculosa). In bird's eggs, the internal cells of the blood island start producing hemoglobin (gas-carrying component of blood) and become the first red blood cells (erythrocytes) as early as the second day of incubation.

The outer cells of the blood islands develop into an endothelial layer and form a network of blood vessels covering part of the surface of the yolk sac. The network acquires a connection to the vitelline veins and vitelline arteries (the latter being branches of the dorsal aorta); thus the blood corpuscles formed in the blood islands can enter the general blood circulation.

At later stages of embryogenesis, blood-cell formation shifts from the blood islands to the liver and, still later, to the bone marrow.

The lymphatic system, in a manner similar to the blood vessels, develops by the local aggregation of connective tissue to form lymphatic vessels.

Reproductive Organs

In considering the development of reproductive organs, distinctions must be made between: (1) the origin of sex cells (gametes), (2) the origin and differentiation of the sex glands, or gonads (ovaries and testes), and (3) the origin and development of the supporting parts of the reproductive system (e.g., genital ducts, copulatory organs).

The germ (germinal) cells, which eventually give rise to the gametes, are often segregated from the somatic, or body, cells at a very early stage—during cleavage and before the subdivision of the embryo into ectoderm, mesoderm, and endoderm. In the invertebrate nematodes, the very first of these primordial germ cells is identifiable after as few as five divisions of the egg cell. The germ cell retains the large chromosomes present in the fertilized egg; in the somatic cells the chromosomes become fragmented. Subsequently, the single germ cell gives rise, by mitotic divisions, to all the gametes in the gonad.

In vertebrates, primordial germ cells arise outside the gonads, but they cannot be distinguished in early cleavage stages. In amphibians, cytoplasm at the vegetal pole, rich in ribonucleic acids, becomes incorporated into a number of cells, which, during cleavage and gastrulation, lie among the yolky endoderm cells. Later they migrate into the mesodermal layer and become incorporated into the rudiments of the gonads. In higher vertebrates, primordial germ cells can be recognized in the extra-embryonic endoderm of the yolk sac. In mammals, these cells subsequently migrate into the mesoderm and are located in the gonad rudiments. The mouse embryo, for example, originally has fewer than 100 primary germ cells; during their migration, however, their numbers increase as a result of repeated divisions, to 5,000 or more in the gonads.

Although the primordial germ cells either may appear before the separation of germinal layers or be found originally in the endoderm, the gonads are invariably of mesodermal origin. In vertebrates, the first trace of gonad development is a thickening of the coelomic lining on either side of the dorsal mesentery and medial to the kidney rudiments. The thickening, elongated anteroposteriorly, is known as the germinal ridge. The ridge protrudes into the coelomic cavity, and the fold of thickened epithelium becomes filled with mesenchyme. At this stage the primordial germ cells invade the rudiments of the gonads and become associated with the somatic cells of the germinal ridge. In the functionally differentiated gonads, only the actual gametes and their predecessors (spermatogonia and oogonia) are derived from the primary germ cells; the supporting cells are somatic cells of local mesodermal origin. In the ovaries, the follicle cells surrounding and nourishing the young egg cells (oocytes) are of somatic origin, as are also the connective tissue and blood vessels of the gonad. In the testes, supporting elements called Sertoli cells are somatic cells, as are the interstitial cells, which are scattered between the sperm-carrying tubules of the testes and believed to be the source of male hormones.

In the early stages of their development—even while the gonad rudiment is being invaded by primordial germ cells—the female and male gonads are in an indifferent stage. Only later does tissue differentiation of the gonads begin and male or female gonadal developments proceed.

The genital ducts, by which the eggs and sperm are carried away from the gonads, are, in vertebrates, linked with the excretory system. In the male, the seminiferous tubules connect with the nephric tubules of the mesonephros, and the sperm are carried to the exterior by way of the mesonephric duct. In males of lower vertebrates, the mesonephric duct thus serves as a channel both for urine and for sex cells. In amniotes the development of the metanephros as the urine excreting organ has freed the mesonephric duct to carry products associated only with reproduction. In the female, a separate duct, the paramesonephric duct (Müllerian duct), develops beside the mesonephric duct. At its anterior end it utilizes the funnels of the pronephric tubules as its entrance (ostium). The paramesonephric duct develops initially in both female and male embryos. The ducts remain in an indifferent stage longer than the gonads. Eventually the sex hormones produced by the differentiating gonads cause a corresponding differentiation of the ducts. The mesonephric ducts, which become reduced in female embryos, remain in male embryos as ducts for conveying sperm (ductus deferens). The paramesonephric ducts, on the other hand, degenerate in male embryos but become the oviducts in female embryos. In mammals, the terminal portions of the paired oviducts differentiate as two uteri, which, in primates and man, fuse to form a single uterus.

In all terrestrial vertebrates except the placental mammals, the genital ducts, as well as the ducts of excretory organs, open into the cloaca. In mammals, however, the cloaca becomes subdivided into a dorsal part, which conveys the feces, and a ventral part, which receives excretory and genital products. In male mammals the excretory and genital ducts remain connected, having the urethra as their common outlet; in females the urethra serves only for the passage of urine and the uterus opens separately by means of the vagina. In nearly all vertebrates, the male nephric duct is utilized in some degree for the conduction of sperm.

Copulatory organs have developed independently in several groups of vertebrates having internal fertilization. The penis in mammals develops from an outgrowth called the genital tubercle, located at the anterior edge of the urinogenital orifice. The tubercle is laid down in a similar way in embryos of both sexes, and the region of the urinogenital orifice remains in an indifferent state even longer than do the genital ducts. In a comparatively late stage of embryonic life the genital tubercle of male embryos encloses the urethral canal and becomes the penis; in female embryos it remains small and becomes the clitoris.

Endodermal Derivatives

The Alimentary Canal

The alimentary canal is the chief organ developing from endoderm. The way it forms depends on the type of egg cleavage. In eggs with holoblastic (complete) cleavage, after gastrulation the invaginated mass of endoderm lines the archenteron, the cavity of which becomes the alimentary canal, or gut. In eggs with meroblastic (partial) cleavage—and also in mammals (despite their complete cleavage)—the endoderm is produced in the form of a sheet lying flat over the yolk-sac cavity. Subsequently, folds of endoderm and splanchnic mesoderm appear—first anteriorly, then laterally, and lastly posteriorly—and sink, converging ventrally under the embryo and cutting off the future gut cavity from the cavity of the yolk sac. The most anterior and posterior portions of the gut separate, but the middle part remains in open communication with the yolk sac throughout embryonic life, eventually becoming reduced to the yolk stalk, which passes through the umbilical cord.

The alimentary canal of vertebrates becomes differentiated into the oral cavity, pharynx, esophagus, stomach, and intestine. Whether derived from an archenteron or formed by folding of the endodermal sheet, the canal initially does not possess an opening at its anterior end. This is also the case in some lower chordates and echinoderms, which are grouped together with vertebrates as the Deuterostomia, or animals with secondary mouths.

In vertebrates, a mouth forms by a rupture at the anterior end, where the endoderm is in contact with ectoderm. The ectoderm of the future mouth region becomes depressed, forming a mouth invagination, or stomodaeum. The ectodermal and endodermal layers separating the cavity of the stomodaeum from the gut fuse to form the oropharyngeal membrane, which thins and ruptures, providing free passage from the exterior to the gut. Because of its mode of origin, the oral cavity is in part lined by ectoderm and in part by endoderm, the two parts becoming indistinguishable. Before the oropharyngeal membrane ruptures, however, a small pocket forms on the dorsal side of the stomodaeal invagination. This, the rudiment of the anterior lobe of the hypophysis, becomes apposed to the ventral surface of the diencephalon and loses its connection with the mouth cavity.

The anal opening in some exceptional cases (urodele amphibians) is derived directly from the blastopore, which persists as a narrow canal after completion of gastrulation. In other vertebrates, however, the anus develops either near the location of the former blastopore or in a corresponding region at the posterior end of the embryo, where the last remnants of mesoderm migrated to the interior. It is thus claimed that the anus in vertebrates is derived, directly or indirectly, from the blastopore. The mode of formation of the opening is somewhat similar to that of the mouth. A slight invagination of the ectoderm occurs, and a cloacal membrane forms, separating the ectodermal invagination from the gut cavity. The membrane ruptures later to provide the anus.

Pharynx and its Outgrowths

The anterior portion of the endodermal gut, lying immediately posterior to the mouth cavity, expands laterally as the pharynx. The lateral pockets of the pharyngeal cavity, called the pharyngeal pouches, perforate the mesodermal layer, reach the ectoderm, and break through to form pharyngeal, or gill, clefts. In fishes and larvae of amphibians, these clefts develop gills and become respiratory organs. Pharyngeal pouches develop in the early embryos of all vertebrates, including the air-breathing terrestrial reptiles, birds, and mammals. The number of pouches has been reduced in the course of evolution from six or more to four in tetrapods, and the posterior pouches may not actually break through.

The consistent development of pharyngeal pouches and clefts indicates their importance in vertebrate development. Many parts of the vertebrate body are derived from, or dependent on, the pharyngeal pouches; for example, the aortic arches—the most important blood vessels of a vertebrate—develop between successive pharyngeal pouches. Skeletal visceral arches also occur between consecutive pharyngeal pouches (they do not develop if the pharyngeal pouches are prevented from developing). In adult terrestrial vertebrates, parts of the visceral arches are transformed into the hyoid apparatus, supporting the tongue, the auditory ossicles, and parts of the larynx and trachea. Furthermore, some of the material of the pharyngeal pouches is utilized for the formation of the parathyroid glands and the thymus; the former are indispensable glands of internal secretion, and the latter are a source, in mammals, of cells that produce antibodies. The pharynx also produces the rudiment of the thyroid gland as a ventral outgrowth.

Liver, Pancreas and Lungs

Three additional important organs develop from the endoderm: the liver, the pancreas, and the lungs. The liver develops as a ventral outgrowth of the endodermal gut just posterior to the section that eventually will become the stomach. Initially, the liver takes the form of a tubular gland, but it soon acquires a close relationship to the blood sinuses and capillaries, forming lobules around blood vessels rather than around glandular ducts. The pancreas develops from three independent rudiments: two ventral ones, formed just posterior to the liver rudiment, and a dorsal one. The ventral and the dorsal rudiments fuse in most vertebrates to form one organ with a complicated system of ducts opening into the duodenum, a portion of the small intestine. The lungs develop from a ventral hollow outgrowth of the gut, which is located just posterior to the pharyngeal region; the outgrowth branches into a right and left trunk that grow posteriorly beside the esophagus and then expand into hollow sacs, in lower terrestrial vertebrates, or into a system of tubes, in birds and mammals.

The endodermal parts of the alimentary system are, along their entire length, encased by the splanchnic mesoderm of the lateral plates. The coelomic cavities of the right and left sides fuse ventral to the gut but remain separated dorsally by their respective walls, which form the dorsal mesentery—a double membrane by which the gut is suspended from the dorsal side of the body cavity and through which blood vessels and nerves reach the gut. The layer of splanchnic mesoderm next to the endoderm produces the connective tissue and muscular layers of the gut. During development of the glands of the alimentary canal (e.g., pancreas, salivary glands), the mesoderm forms a connective tissue capsule around the branching tubules of the gland. The development of the tubules is dependent on this mesodermal capsule and cannot proceed without it.

Postembryonic Development

The period in the ontogenesis of multicellular animals that follows the period of embryonic development and usually ends with the onset of sexual maturity and, in most animals, the cessation of growth. Postembryonic development begins after the emergence of the embryo from the egg and embryonic membranes, when the organism becomes capable of active feeding and locomotion. Upon transition to the postembryonic state the organism either immediately possesses the principal pubertal morphological characteristics (direct development) or essentially differs from the pubertal form, in which case the larva that hatches out of the egg must undergo a metamorphosis before it reaches its adult state. Growth continues during the period of postembryonic development and further organogenesis and histogenesis occur. The functions of the developing organism become more complex; establishment of the final proportions of the body is especially characteristic.

In some animals postembryonic development constitutes the greater part of their life span. Among insects, in a number of cicadas the larva lives 17 years, while the pubertal insect lives one summer; the mayfly larva lives to three years, but often in its pubertal form the mayfly lives only one day.

The necessity for defense against enemies and active food procurement are ensured during postembryonic development by a number of adaptations that make the animal self-sufficient, sometimes for its entire life span. With indirect, or larval, development, the animal, having larval organs, is self-sufficient only during the postembryonic period.

References

- Animal-development, science: britannica.com, Retrieved 10 April, 2019
- Animal-development, growth-and-reproduction: biology.gatech.edu, Retrieved 12 May, 2019
- Embryonic-induction, animal-development, science: britannica.com, Retrieved 14 June, 2019
- Postembryonic+Development: thefreedictionary.com, Retrieved 16 July, 2019

Permissions

All chapters in this book are published with permission under the Creative Commons Attribution Share Alike License or equivalent. Every chapter published in this book has been scrutinized by our experts. Their significance has been extensively debated. The topics covered herein carry significant information for a comprehensive understanding. They may even be implemented as practical applications or may be referred to as a beginning point for further studies.

We would like to thank the editorial team for lending their expertise to make the book truly unique. They have played a crucial role in the development of this book. Without their invaluable contributions this book wouldn't have been possible. They have made vital efforts to compile up to date information on the varied aspects of this subject to make this book a valuable addition to the collection of many professionals and students.

This book was conceptualized with the vision of imparting up-to-date and integrated information in this field. To ensure the same, a matchless editorial board was set up. Every individual on the board went through rigorous rounds of assessment to prove their worth. After which they invested a large part of their time researching and compiling the most relevant data for our readers.

The editorial board has been involved in producing this book since its inception. They have spent rigorous hours researching and exploring the diverse topics which have resulted in the successful publishing of this book. They have passed on their knowledge of decades through this book. To expedite this challenging task, the publisher supported the team at every step. A small team of assistant editors was also appointed to further simplify the editing procedure and attain best results for the readers.

Apart from the editorial board, the designing team has also invested a significant amount of their time in understanding the subject and creating the most relevant covers. They scrutinized every image to scout for the most suitable representation of the subject and create an appropriate cover for the book.

The publishing team has been an ardent support to the editorial, designing and production team. Their endless efforts to recruit the best for this project, has resulted in the accomplishment of this book. They are a veteran in the field of academics and their pool of knowledge is as vast as their experience in printing. Their expertise and guidance has proved useful at every step. Their uncompromising quality standards have made this book an exceptional effort. Their encouragement from time to time has been an inspiration for everyone.

The publisher and the editorial board hope that this book will prove to be a valuable piece of knowledge for students, practitioners and scholars across the globe.

Index

www.ingramcontent.com/pod-product-compliance
Lightning Source LLC
LaVergne TN
LVHW080648210926
848485LV00003B/345